L'Organisation de Coopération et de Développement Économiques (OCDE), qui a été instituée par une Convention signée le 14 décembre 1960, à Paris, a pour objectif de promouvoir des politiques visant :

— à réaliser la plus forte expansion possible de l'économie et de l'emploi et une progression du niveau de vie dans les pays Membres, tout en maintenant la stabilité financière, et contribuer ainsi au développement de l'économie mondiale;
— à contribuer à une saine expansion économique dans les pays Membres, ainsi que non membres, en voie de développement économique;
— à contribuer à l'expansion du commerce mondial sur une base multilatérale et non discriminatoire, conformément aux obligations internationales.

Les Membres de l'OCDE sont : la République Fédérale d'Allemagne, l'Australie, l'Autriche, la Belgique, le Canada, le Danemark, l'Espagne, les États-Unis, la Finlande, la France, la Grèce, l'Irlande, l'Islande, l'Italie, le Japon, le Luxembourg, la Norvège, la Nouvelle Zélande, les Pays-Bas, le Portugal, le Royaume-Uni, la Suède, la Suisse et la Turquie.

L'Agence de l'OCDE pour l'Énergie Nucléaire (AEN) a été créée le 20 avril 1972, en remplacement de l'Agence Européenne pour l'Énergie Nucléaire de l'OCDE (ENEA) lors de l'adhésion du Japon à titre de Membre de plein exercice.

L'AEN groupe désormais tous les pays Membres européens de l'OCDE ainsi que l'Australie, le Canada, les États-Unis et le Japon. La Commission des Communautés Européennes participe à ses travaux.

L'AEN a pour principaux objectifs de promouvoir, entre les gouvernements qui en sont Membres, la coopération dans le domaine de la sécurité et de la réglementation nucléaires, ainsi que l'évaluation de la contribution de l'énergie nucléaire au progrès économique.

Pour atteindre ces objectifs, l'AEN :

— *encourage l'harmonisation des politiques et pratiques réglementaires dans le domaine nucléaire, en ce qui concerne notamment la sûreté des installations nucléaires, la protection de l'homme contre les radiations ionisantes et la préservation de l'environnement, la gestion des déchets radioactifs, ainsi que la responsabilité civile et les assurances en matière nucléaire ;*
— *examine régulièrement les aspects économiques et techniques de la croissance de l'énergie nucléaire et du cycle du combustible nucléaire, et évalue la demande et les capacités disponibles pour les différentes phases du cycle du combustible nucléaire, ainsi que le rôle que l'énergie nucléaire jouera dans l'avenir pour satisfaire la demande énergétique totale ;*
— *développe les échanges d'informations scientifiques et techniques concernant l'énergie nucléaire, notamment par l'intermédiaire de services communs ;*
— *met sur pied des programmes internationaux de recherche et développement, ainsi que des activités organisées et gérées en commun par les pays de l'OCDE.*

Pour ces activités, ainsi que pour d'autres travaux connexes, l'AEN collabore étroitement avec l'Agence Internationale de l'Énergie Atomique de Vienne, avec laquelle elle a conclu un Accord de coopération, ainsi qu'avec d'autres organisations internationales opérant dans le domaine nucléaire.

FOREWORD

The process of siting radioactive waste repositories in deep geological formations is a complex one, that includes a number of phases. Site selection usually begins with a general review of available geologic information and with reconnaissance work in the field. More detailed investigations follow, such as geological mapping, geophysical surveys and exploratory drilling. Once a site has been tentatively selected, it must be characterised in sufficient detail to prove that it can actually accommodate a radioactive waste repository with the required level of safety. The site characterisation process includes extensive field work and can go as far as sinking a shaft and excavating one or more experimental rooms.

A number of NEA Member countries are engaged in studies to select, evaluate and characterise sites for disposal of radioactive wastes in mined geologic repositories. Other countries have not yet begun this process, but are planning activities in the near future. Although there is general agreement on the broad aspects of the repository siting process, countries are just now addressing the various aspects in detail. Certainly, more international exchange of views on the details of the site selection and characterisation process would enhance understanding of differing approaches. The way they are modified to meet national or local conditions might also be very instructive for all parties. Finally the discussion of how public acceptance is being handled in the different countries might lead to some badly needed progress in this difficult area.

The objectives of this workshop were :

- to review the approaches used in NEA Member countries for the siting of repositories ;
- to consider what information is required to determine whether a particular location is suitable as a disposal site ;
- to review the technical means that can be used to obtain the necessary information ;
- to discuss possible approaches for effectively communicating to the public the results of the site characterisation process.

The papers and discussions at the meeting are recorded in these proceedings which provide a synthesis of the present thinking in this field.

AVANT-PROPOS

Le choix des sites des dépôts de déchets radioactifs dans des formations géologiques profondes est un processus complexe qui comprend plusieurs phases. On commence habituellement par passer en revue les données géologiques disponibles et par faire des travaux de reconnaissance sur le terrain. Viennent ensuite des recherches plus détaillées, telles que l'établissement de cartes géologiques, des levées géophysiques et des forages d'exploration. Lorsqu'un site est envisagé comme possible, il faut en déterminer les caractéristiques de façon suffisamment précise pour démontrer qu'il peut effectivement servir de réceptacle pour l'isolement de déchets radioactifs tout en offrant le degré de sûreté requis. Le processus de caractérisation du site entraîne d'importants travaux qui peuvent aller jusqu'au fonçage d'un puits et à l'excavation d'une ou plusieurs salles expérimentales au niveau choisi pour l'évacuation.

Dans plusieurs pays Membres de l'AEN, des études sont en cours, qui visent à choisir, évaluer et caractériser des sites pour l'évacuation de déchets radioactifs dans des dépôts géologiques creusés à cet effet. D'autres pays n'en sont pas encore à ce stade mais envisagent des activités de cette nature pour l'avenir proche. Si tous les pays sont d'accord sur les principes généraux du processus de choix des sites des dépôts, ils commencent tout juste à s'intéresser aux détails de la question. Le développement des échanges de vues au niveau international sur les détails du choix des sites et sur le processus de caractérisation ne peut évidemment que contribuer à faire mieux comprendre les méthodes choisies ou envisagées dans chaque pays. De plus, la façon dont ces méthodes varient en fonction des conditions nationales ou locales peut également être très instructive pour tous les intéressés. Enfin, l'examen de la manière dont chaque pays aborde le problème de l'acceptation par le public permettrait peut-être de réaliser des progrès notables dans ce domaine délicat.

Les objectifs de cette réunion de travail étaient les suivants :

- passer en revue les méthodes utilisées dans les pays Membres de l'AEN pour le choix des sites des dépôts ;

- examiner la nature des données nécessaires pour déterminer si un emplacement particulier convient comme site d'évacuation ;

- passer en revue les moyens techniques qui peuvent être utilisés pour obtenir ces données ;

- débattre des méthodes susceptibles d'être utilisées pour informer efficacement le public des résultats du processus de caractérisation des sites.

Les communications et les échanges de vues intervenus au cours de la réunion sont présentés dans ce compte rendu qui constitue une synthèse des concepts actuels dans ce domaine.

TABLE OF CONTENTS

TABLE DES MATIERES

SESSION I - SEANCE I

Chairman - Président : M. A. BARBREAU (France)

SESSION II - SEANCE II

Chairman - Président : Dr. M.F. THURY (Switzerland)

SESSION III - SEANCE III

Chairman - Président : Mr. L.J. ANDERSEN (Denmark)

SESSION IV - SEANCE IV

Chairman - Président : Mr. M.S. BEDINGER (United States)

SESSION V - SEANCE V

Chairmen - Présidents *Dr. F. GERA (Italy)*
Mr. J.D. MATHER (United Kingdom)

SESSION I

Chairman - Président

A. BARBREAU

(France)

SEANCE I

SITE SELECTION IN THE NATIONAL WASTE TERMINAL STORAGE PROGRAM: AN OVERVIEW OF TECHNICAL AND POLITICAL ASPECTS OF THE PROBLEM

W. E. Newcomb, S. S. Smith, and O. E. Swanson

May 19, 1981

ABSTRACT

The Department of Energy's mined geologic repository program is seeking early access to deep-target horizons. The program is focusing on three sites in three lithologies: basalt, tuff, and salt. The scientific and technical program is now focusing on what earth-science data are required to adequately characterize any given repository site from both technical and regulatory perspectives. To that end, a number of questions are posed, the answers to which are necessary for program planning, management, and execution with respect to earth science information.

The U.S. Department of Energy is conducting a national program, the goal of which is to provide mined geologic repositories for high-level radioactive waste. To accommodate the volume of waste expected to be produced by the United States, it is the intention of the Department of Energy to select an as-yet-undetermined number of sites across the nation. The siting portion of the program has both the highest visibility and the greatest sensitivity.

The Department of Energy's decision to select sites to be licensed can only be achieved, technically and politically, by completing several step-wise activities. These activities, which can be thought of as discrete stages leading to site selection, are as follows:

- National screening to identify places of promise for characterization
- Characterization and documentation of these places to identify potential sites
- Detailed characterization and documentation of potential sites
- Site selection.

These stages are intended to provide a technically conservative, politically sensitive, stepwise progression toward the site-selection goal. The process also makes the program cost-effective; because the search focuses on progressively smaller units of land, at each successive stage, exploration funds are expended at places having greater and greater likelihood of providing adequate repository sites.

The success of this program very largely hinges on the technical adequacy of the characterization and documentation steps, combined with the sociopolitical processes by which the Department of Energy uses the technical information to support its eventual selection decisions. Therefore, sociopolitical issues and concerns in the characterization and documentation program must be considered before and during characterization, in order that the selection decision be supportable.

Our immediate concern is collection of all data required for site characterization. This is clearly necessary for technical reasons that are generally obvious to the scientific community but it is equally important to allow the technical work to support the sociopolitical process, which is intended to give the public confidence that its concerns have been addressed. The scientific requirements must be presented in a comprehensive manner to facilitate planning, execution, and integration of the technical work.

In the United States' program, the Department of Energy is now emphasizing achievement of early access (in the mid-1980's) to at-depth target horizons in several lithologies. These efforts are focused on the following projects:

- BASALT WASTE ISOLATION PROJECT: The Basalt Waste Isolation Project is evaluating the Department of Energy's Hanford Reservation, in the center of the Pasco Basin, Washington, to determine whether it contains a suitable site for a repository in basalt. Geologic and hydrologic investigations have been under way since 1977, representing a continuation of investigations conducted between 1968 and 1972. Investigations of the Hanford reservation to date suggest that it has adequate geologic characteristics to serve as a repository site. The location and movement of water in the unconfined and upper basalt confined aquifers, as well as questions about the location and movement of water in interbeds and interflows of deeper (Wanapum and Grand Ronde) basalts are being addressed. The tectonic conditions of the area are under investigation, and the area appears to be sufficiently stable for siting a repository.

- NEVADA NUCLEAR WASTE STORAGE INVESTIGATIONS: These investigations are evaluating the suitability of tuff at the Department of Energy's

Nevada Test Site in Nevada. Current data and understanding indicate that tuffaceous rocks on the test site have potential for isolating radioactive wastes. Identification of a site with suitable geologic and hydrologic characteristics is the primary focus of exploration. The likelihood and the consequences of future tectonic events are being analyzed to determine whether the tectonic setting would preclude waste disposal at the test site. Argillite, granite, alluvium, and tuff have been considered as potential host rocks, but current exploration efforts in the southwestern part of the Nevada Test Site are directed at places underlain by tuff.

At present, only the tuff at Yucca Mountain is being explored. Field mapping, core drilling, and geophysical surveys are in progess to assess the extent to which satisfactory geological conditions exist at Yucca Mountain. A 6,000-ft core and hydrologic test hole has been drilled into the study area, and additional borings are currently underway. The results will be compared and correlated with data from a 2,500-ft hole drilled earlier. The water-bearing properties of inferred fracture zones in Yucca Mountain are being evaluated by hydrologic testing and geophysical surveys. The geochemical and rock mechanical properties of tuff in the Nevada Test Site are also under investigation in the laboratory.

OFFICE OF NUCLEAR WASTE ISOLATION SALT INVESTIGATIONS: DOE's program for investigating salt as a repository medium is under the direction of the Office of Nuclear Waste Isolation. Investigations are currently under way at three separate localities in the United States: the Gulf Coast salt dome basins, the Permian Basin, and the Paradox Basin. Current strategy calls for a prioritization to be made between the bedded salts of the Permian and the Paradox Basin within the next 12 months. Following that, investigation would proceed on the prioritized bedded-salt region, in parallel with work on the Gulf Coast salt domes. At a later date, probably within approximately 2 years, a decision would be made as to which of the remaining salt hosts, bedded or dome, would receive priority and eventually be compared with the site in basalt (Hanford Reservation) and the site in tuff (Nevada Test Site).

Gulf Coast Salt Dome Basins: The Gulf coastal plain of Texas, Louisiana, and Mississippi and adjacent offshore areas, contains more than 500 salt domes, 263 of which exist on land. The Department of Energy identified 8 of these salt domes for closer scrutiny several years ago; two in the State of Louisiana, three in the State of Mississippi, and three in the State of Texas. One of the Texas domes was dropped from consideration two years ago because of uncertainties related to earlier solution mining.

Salt-dome studies have stressed evaluation of tectonic and hydrologic stability. Efforts to assess hydrologic stability are centered on determining the reasons for the apparent resistance of salt masses to dissolution. Investigations to date have included regional hydrologic studies of the three sedimentary basins in which the domes occur, as well as dome-specific geologic and hydrologic studies. Aquifers have been investigated by borehole pump tests to depths as great as two kilometers in each basin.

Geologic studies have included regional and dome-specific field mapping with emphasis on evaluating remote sensing data, geophysical well logs, and seismic reflection surveys. Dome boundaries have been defined by gravity surveys, in conjunction with high-resolution seismic reflection surveys and boreholes. Studies of tertiary and older strata suggest that salt movement in the Gulf Coast's interior region ceased perhaps as long as 30 million years ago.

Paradox Basin: The Paradox Basin underlies southeastern Utah and southwestern Colorado. Studies are currently focused on the Utah portion of the basin which is underlain by bedded salt. The geological setting of the Paradox Basin has been characterized and evaluated in order to select four study areas that are now being investigated. These four study areas lie in Utah and are: Salt Valley, Lisbon Valley, Gibson Dome, and Elk Ridge. Factors considered in identifying these places include: depth to salt, thickness of salt, location of mapped faults, location of urban places, location of designated lands (e.g., national parks), and mineral resources. Current field investigations include surface and subsurface geophysical, geochemical, hydrogeological, and geological studies. Results of companion environmental investigations have been and are being factored into the decision process.

Field investigations of Salt Valley, a salt diapir, have been completed. Investigations included a continuously cored 4,100-ft borehole, hydrologic testing, a seismic refraction survey, crosshole vertical seismic survey, and surface electrical and electromagnetic surveys.

Strata in the Gibson Dome and Elk Ridge areas are nearly horizontal. Field work in both areas has included mapping exposed stratigraphic marker beds, continuous coring and testing of holes bottomed approximately 500 feet below the base of the Paradox member of the Hermosa formation, field checking of mapped lineaments, hole-to-surface DC resistivity surveys, surface electromagnetic surveys, and excavation of soil pits as part of an examination of Quaternary history. A seismic reflection line is planned for the Elk Ridge area. A temporary 24-station microearthquake network, operated continuously for six months, was reduced to a 12-station network to monitor an area that includes a diapiric anticline and the intersection of northeast and northwest trending structural features.

Lisbon Valley, a nondiapiric salt anticline, is in an area which has been extensively surveyed by the oil industry in the past. Plans for evaluation of this area are to rely on these existing data and other literature, combined with a limited amount of aerial and ground reconnaissance to evaluate the area's suitability.

Permian Basin: The Permian Basin comprises several sedimentary sub-basins in which evaporites accumulated. The basin includes the western parts of Kansas, Oklahoma, and Texas, and the eastern parts of Colorado and New Mexico. Since Permian time, the basin has been relatively stable tectonically, although some parts of it have been tilted and warped. The Palo Duro and Dalhart subbasins of the Permian Basin were identified as areas with potential for repository siting by screening the known information on the distribution of bedded salt within the Permian Basin. These subbasins are located largely in the Texas Panhandle, an area typified by an almost featureless plain, dissected by headward cutting streams, and underlain by nearly horizontal Mezozoic and Cenozoic sediments.

Detailed investigations of the Permian Basin began only in the last year. The currently available data are preliminary. Specific questions pertaining to hydrology, tectonics, geology, and resource evaluations are to be the subjects of future investigation. Eight thousand feet of salt-bearing core have been recovered in the past three years, and plans are for two additional exploratory holes within the next 12 months. These holes will address both stratigraphic concerns and hydrologic (e.g., salt dissolution) issues. Quantitative data on resource potential, climatic history, erosion potential, and salt dissolution have been and are continuing to be collected to assist in predicting the long-term geomorphic integrity of the Texas Panhandle and the likelihood for possible human intrusion.

The goal of our next few years of investigation is to identify, from among the foregoing areas, three places, in three lithologies, at which the Department of Energy will sink an exploratory shaft to total depth. When shaft construction has been completed and at-depth investigations have been carried out, demonstrating that the site(s) continue to appear suitable, one of the three places will be chosen for construction of a research and development facility.

We are in the process of developing an approach to simultaneously accommodate the Department of Energy's new emphasis and to address the technical issues related to siting in a conservative, stepwise manner. We have begun to think of site characterization in two general senses:

(1) Surface-based characterization, during which we seek data to answer the following questions:

(a) What are the technical issues related to site suitability?

(b) Which issues can be resolved, or at least bounded, using surface-based characterization techniques?

(c) What is the priority for resolving these issues?

(d) Given (a-c), what data must be collected to allow the Department of Energy to be reasonably assured that a site appears adequate to proceed with an exploratory shaft to total depth?

(e) What techniques or tests will best provide the required data?

(2) Subsurface (in situ) characterization, during which we will seek data to answer the following questions:

(a) What technical issues remain to be resolved because surface-based techniques cannot adequately provide the required data?

(b) What technical issues could be merely bounded by surface-based characterization?

(c) Given (a) and (b), what new data are required to resolve the outstanding issues and in what order of priority must they be resolved?

(d) What data are required to confirm the bounding values collected using surface-based techniques?

(e) What techniques or tests will best provide the required data?

To satisfy the technical requirements, and to support the sociopolitical process, each question must be addressed comprehensively for each aspect of the repository system. Comprehensive answers significantly aid in planning these intricate and interdependent investigations for a repository site, making clear commonalities of need or unique requirements.

As a basis for comprehensively addressing site characterization, we have subdivided the site characterization task into seven categories within which data are required. These categories are:

- Geology/geohydrology
- Seismicity/seismic design
- Repository design/construction
- Waste package design
- Repository sealing

- Geochemistry

- Performance assessment.

A preliminary draft document has resulted from our considerations and is constructed to provide comprehensive guidelines on required earth-science-related data for geologic repositories. Smith and Swanson will discuss explicit examples from this document later today, concentrating on geologic and geohydrologic aspects of site characterization.

We believe that when the document has been reviewed, revised, and made final, it will provide a valuable technical baseline for both scientists and management. In addition, as an informational document, it will support the sociopolitical process by allowing interested third parties to examine in substantial detail the elements of science and technology which constitute the Department of Energy's characterization of potential repository sites in the United States. It thus can serve as a vehicle to stimulate discussion of technical concerns among parties having distinct and even conflicting points of view regarding geologic disposal of radioactive waste. While its immediate value will be mostly directed to the technical and managerial portion of the U.S. waste management program, it may well have even greater value, in the long-term, in fostering discussion and understanding among those who must make decisions, in the sociopolitical arena, on the basis of recommendations by technical and scientific bodies.

A specific arena in which this document will, we believe, prove very useful in communication with both political and technical bodies representing the various states in which characterization may be undertaken. The Department of Energy is making strong efforts to involve state agencies substantively in the siting process by inviting their active participation in the investigations and by keeping them fully informed of all related activities. These efforts, referred to as "Consultation and Concurrence", are one key to building the sociopolitical basis for eventual site selection. States frequently ask both what will be done and why it will be done. When detailed site characterization begins, this document will be able to answer both questions and to allow states to determine which aspects of the investigations, if any, they might wish to monitor or undertake. Furthermore, the document will allow the states to assess for themselves the completeness and adequacy of the characterization effort.

We believe that one option for focusing discussions in this meeting is to address the technical questions which are posed above. We welcome your individual and collective perspectives on these questions and on other questions which should appropriately be posed as part of the siting process.

DISCUSSION

G. DE MARSILY, France

Is the WIPP site in New Mexico still considered as a potential site by DOE ?

Why is there a difference made between defence waste and commercial waste ?

O.E. SWANSON, United States

The WIPP site is, at present, not being considered as one of DOE's potential site for high level waste disposal.

Defence (TRU) waste repositories will not be licensed under the USNRC as will commercial high level waste repositories. The WIPP site was originally, and is still, considered to be a site for defence transuranic waste only, and as such will remain separately funded until Congress dictates differently.

THE BRITISH APPROACH TO THE DETERMINATION OF THE PRACTICALITY OF LOCATING A RADIOACTIVE WASTE REPOSITORY IN GEOLOGICAL FORMATIONS ON LAND

J.D. Mather, J.H. Black, P.B. Greenwood and B.C. Lintern

Institute of Geological Sciences, Exhibition Road
London, U.K.

The approach adopted in the United Kingdom is outlined in a statement made by the Secretary of State for the Environment on 24th July 1979. This emphasises the need for a programme of geological research involving test drillings to examine fully the properties and characteristics of different geological formations in-situ. The objective is to demonstrate whether or not geological disposal is a practicable option in the U.K. geological environment and, if it is, to make recommendations to Government on which host rocks would be most suitable for further more detailed assessment in the future. The present geological research programme has evolved to try to meet these objectives over a 10 year time scale. The progress of the work is closely related to the ability to obtain planning permission for test boreholes. Experience to date has demonstrated that expression of a research interest in any area leads to grossly misinformed statements by local anti-nuclear groups and the press. The British programme is discussed from the point of view of its progress, the type of information being acquired, the reactions of the public and the presentation of the results.

1. INTRODUCTION

In Britain responsibility for policy on the management of civil nuclear wastes lies with the Secretary of State for the Environment, together with the Secretaries of State for Scotland and Wales. One of the main elements of this responsibility is to ensure that there is adequate research and development on methods of disposal. Prior to the Environment Departments being given this responsibility it lay with the Department of Energy who conducted research through the United Kingdom Atomic Energy Authority (UKAEA). The UKAEA are still involved in this research as contractors to the Department of the Environment (DoE), however, since the beginning of 1980 the main responsibility for conducting geological investigations has lain with the Natural Environment Research Council's Institute of Geological Sciences (NERC/IGS).

The British Government's views on research into the disposal of high level radioactive wastes are set out in a statement made by the Secretary of State for the Environment on 24th July 1979. This statement emphasises the need for a "programme of geological research involving test drillings to examine fully the properties and characteristics of different geological formations in-situ." It "is not a programme for disposing of radioactive waste but is .... research into whether disposal in geological formations is feasible ". "Only when full information is available, and has been properly evaluated, will it be possible to judge whether or not disposal deep underground is an option to be pursued, and, if it is, which of the rocks would be most suitable". "When research has been conducted for about ten years the Government expect to have obtained sufficient information to enable decisions to be taken about the development of demonstration disposal sites underground ..." More recently Parliament was told "The Government must ensure the continuation of a responsible long-term research programme in the United Kingdom into possible methods of disposing of high-level radioactive waste of which geological disposal may be one" (Hansard, May 13, 1981)

The present geological research programme is directed towards meeting these objectives in the required time scale. It was initiated in 1976 when several areas were identified as possibly fulfilling the criteria for the disposal of highly active waste into crystalline rock formations. From these areas it was envisaged that one or two would be selected for detailed assessment. The idea at this early stage of the research was to demonstrate whether, from a geological viewpoint, it would be possible to construct a repository within a particular host rock at a particular site. During the latter part of 1977 an appraisal of the rationale of the research programme resulted in a change of emphasis towards a much larger feasibility study designed to collect the data necessary to build up a coherent view of the potential of various geological environments in the U.K. In addition continuing evolution of the programme has meant that it is now not only concerned with a study of crystalline rocks but involves all potentially appropriate host rock types which occur in the U.K. geological environment.

The present view of the Institute of Geological Sciences (IGS) is that the effectiveness of the geological barrier provided by potentially suitable U.K. geological environments still has to be demonstrated. The geologist is not yet armed with sufficient knowledge to be able to state, except in broad terms, which geological environments would be most suitable and it is data relevant to this issue which must now be obtained. Information is required to relate the primary factor of interest, which is the potential movement of radionuclides through groundwater circulation, to rock mass and rock material parameters such as mineralogy, geochemistry and petrology, thermomechanical properties, joint frequency, the attitude and openness of fractures, regional and local structures and also to other parameters such as topography and drift cover. The overall programme is designed to obtain comprehensive field data from a range of potentially appropriate geological environments. The aims are to test the criteria on which these environments have been chosen, to develop techniques for the collection and assessment of data and to build up an information base such that the most appropriate environments for consideration as repository sites can be defined.

2. SELECTION OF RESEARCH AREAS

Although there is no intention within the present British programme to characterise a particular environment in sufficient detail to prove that it can actually accommodate a radioactive waste repository it is still necessary to select a range of potentially appropriate geological environments for study. The selection of these research environments involves a process comparable to that which might be used to chose a repository site.

The selection process began in the U.K. in 1975 when IGS were commissioned by the UKAEA to define a set of criteria to be used in the identification of areas in the U.K. which were considered to contain geological formations potentially suitable for high-level waste disposal. The published criteria (Gray et al. 1976) were used to identify areas underlain by crystalline igneous and metamorphic rocks, a range of argillaceous rocks, extending from plastic clays through mudstones to brittle slates, and evaporites. Altogether some 16% of the land area of the U.K. was considered to be underlain by potentially suitable rocks.

2.1 Crystalline rock areas

The early work, mentioned in the introduction to this paper, was concentrated on studies in one or two crystalline rock areas and has since evolved into the present assessment programme. On the basis of desk studies and reconnaissance visits 8 crystalline rock research areas have been selected from a short list of 24. The movement of underground water in crystalline rocks is related to the distribution and character of fractures and discontinuities within the rock mass. These are in turn related to the rock type and geological history of the crystalline rocks. Thus the 8 areas selected include a wide range of crystalline rock types and structural settings and incorporate a spectrum of different geomorphological features. The influence of these factors on groundwater movement and hence on the geological barrier which the rocks provide can be examined in detail. The areas are described in detail by Mather et al. (1979).

2.2 Argillaceous rocks

Underground water moves within the more consolidated argillaceous rocks both through fractures and discontinuities and through intergranular pore spaces. The distribution of these different types of movement is related to the geological history of the rocks e.g. to their depth of burial, their involvement in large scale earth movements etc. To a certain extent this geological history is in turn related to the age of the rock. Thus the older, harder argillaceous rocks are generally fractured and jointed, whereas the younger rocks are plastic and will have few discontinuities. Additional parameters which need to be taken into account include the role of relief and other factors, such as depth and the overlying and underlying rocks, in controlling flow within sedimentary basinal environments. Four areas have been selected representing a range of argillaceous rocks to include formations of different ages and evolutionary histories so that a spectrum of hydrogeological environments can be assessed and their influence on the geological barrier provided by argillaceous rocks determined (Mather et al. 1979). It is anticipated that in the future additional argillaceous rock environments will be selected including composite environments where younger plastic argillaceous rocks overlie older fractured rocks at depth.

2.3 Evaporites

Although evaporites are a favoured disposal formation in other countries their potential within the land area of the U.K. is limited. Diapiric structures are not known to occur beneath the U.K. mainland and the bedded salt deposits are relatively thin and interbedded with mudstones. Three areas have been selected (Mather et al. 1979) to represent these rocks in order to assess the integrity of the geological barrier which they might provide.

It should be emphasised that the research areas have been selected using geological criteria and information on land ownership. Non-geological criteria which need to be taken into account at a potential repository site such as access planning constraints etc. have not been considered in selecting the research areas.

3. PROGRESS OF FIELD PROGRAMME

In his statement of 24th July 1979 the Secretary of State announced that "All exploratory work, including test boring, in any area, whether or not on land owned by the Crown or a Government Agency will be the subject of appropriate planning procedures". These procedures involve the completion of an application form and, under certain circumstances, require that the development is advertised in the press and that notices are posted on the site. The application form is submitted to the local district planning authority and a decision on the application may be taken by them or, by agreement, it may be deferred to the County or Regional Authority for a decision. In each case the decision is taken by elected representatives who receive advice from professional officials.

If the decision is positive the work can proceed immediately taking into account any conditions imposed by the planning authority. If a decision against the application is made it is then open to the applicant to submit an appeal to the relevant Secretary of State (either the Secretary of State for the Environment in England, or the Scottish or Welsh Secretaries of State in those countries) who appoints an inspector to hold a local public inquiry to consider evidence presented by both sides and to make a recommendation to him. The ultimate decision then lies with the Secretary of State.

Planning applications were submitted by the UKAEA for drilling work in 3 crystalline rock areas in 1978; one of these applications, at Altnabreac in Caithness in North East Scotland, was granted and drilling took place between November 1978 and May 1979. However, the other applications in Northumberland in north-east England and Ayrshire in south-west Scotland were rejected. Appeals were submitted and local public inquiries were held in Ayrshire during February and March 1980 and in Northumberland during October and November 1980. The results of these appeals are still awaited which means that 3 years after the applications were submitted no final decision has been anounced on whether or not exploratory drilling can go ahead.

Planning applications for work in two argillaceous rock areas were submitted by NERC/IGS in October 1980. Of these one was refused while in the other the planning authority chose not to decide the application - a situation termed a "deemed refusal". In both cases it is now open to IGS to submit an appeal and this is being actively considered at the present time.

The description of the progress of the programme illustrates the considerable delays which may result from the planning process and the long lead-times which are required between formulating an interest in a particular area and gaining permission to work there. Initial visits were paid to the Ayrshire area in August 1976 so that nearly 5 years have elapsed since the first reconnaissance and a decision on whether work can or cannot go ahead is still awaited.

4. PRESENTATION OF FIELD PROGRAMME AND PUBLIC REACTION

A press notice was issued at the same time as the Secretary of State made his statement in July 1979 and listed in a general way a number of areas of interest for research boreholes. Following this announcement the research areas were identified specifically by Mather et al.(1979). The statement was reported factually by the media but did not receive much coverage at the time.

Four of the areas outlined in the statement were selected for more detailed studies with a view to possible subsequent submission of planning applications. These localities were announced in a national press statement released in mid-January 1980, which naturally focussed attention on them and led to considerable attention by the local media but little national coverage. However, at the same time the public inquiry into the Ayrshire application, which had been part of the earlier programme, led to considerable national publicity particularly in Scotland where the inquiry took place. This publicity resulted in a summer of discontent for geologists, particularly in Scotland where any geological activity was seen as contributing to the radioactive waste disposal programme and was therefore the subject of harrassment. Even parties of student geologists were implicated as they were said to be being used by the nuclear industry, without their knowledge, to provide data for the programme.

Government geologists involved in the programme and the geological profession as a whole reacted to the rather unaccustomed glare of publicity and, amid a lively exchange of correspondence in the local and national press in Scotland, most geological fieldwork unrelated to the radioactive waste research managed to go ahead. However, the message was well and truely brought home of how the activities of a small group of geologists working on the radioactive waste disposal research programme can have a major effect on the whole geological profession.

There is certainly a myth surrounding high-level radioactive waste and its disposal. The properties attributed to this waste by many well-intentioned but misinformed people means that it engenders fear out of all proportion to its capacity to do harm. The expression of a research interest in any particular area leads to the fueling of this fear and to grossly misinformed statements by local anti-nuclear groups and the press. The fact that the object of the research is to demonstrate the feasibility or otherwise of geological disposal is generally overlooked and the area immediately becomes a chosen site for "waste dumping" in the public mind. In these circumstances it is not suprising that local elected representatives who have to decide planning applications, vote to prevent test drilling taking place in their area even when they are advised by their professional officers that there are no sound reasons under the planning laws for doing so.

It is apparent that within the United Kingdom the method of presentation of the research programme has not convinced the public of either the need for or the purpose of the research. In an attempt to allay the misconceptions about the programme, the Institute of Geological Sciences provided a much longer information package with the last planning applications which were submitted in October 1980.

A press announcement was issued immediately before the applications were submitted and 15 pages of explanatory notes accompanied the applications themselves. These notes described the independent position of NERC and its component bodies such as IGS. They went on to discuss government policy on radioactive waste disposal research and to describe the objective of the planned research on the particular area. The reasons for the choice of that area were emphasised and the notes concluded with a description of the proposed drilling operation. Copies of these notes were also distributed to local parish councils and to any member of the public who made an enquiry concerning the application. In addition representatives of IGS and the Department of the Environment addressed local meetings to explain the programme. Local authorities were encouraged to organise meetings if there was a demand but no attempt was made by IGS to convene local meetings. This general approach led to a much more reasoned debate on the applications but the end result was still a rejection or a refusal to decide them one way or the other.

5. INFORMATION RESULTING FROM THE FIELD PROGRAMME

The object of the IGS field programme is to gain information on the geological barrier provided by a range of potentially suitable host rocks. Part of the programme concentrates on the movement of groundwater within these host rocks and on the factors which control this movement. The natural groundwater flow systems will be perturbed by the construction of a repository and the rock-heating and cooling cycle which will follow from waste emplacement. The original system will then re-establish itself over the long period during which the temperature returns to ambient. The analysis of the perturbations is the responsibility of other agencies and the IGS work is related to the study of the natural systems.

The only research site where drilling work has been initiated so far is at

Altnabreac in Caithness, north-east Scotland. Hydrological and near-surface hydrogeological investigations have been carried out using surface water data, spring data and information from 24 shallow boreholes to around 40m below surface. A packer system has been developed to measure hydraulic conductivity in the range 10^{-6} to 10^{-12} m/sec. The system differs from the normal practice in that all measurements are achieved by abstraction of water rather than injection. In so doing, good samples of water for chemical and isotope analysis have also been obtained without the problem of having introduced extraneous water into the groundwater system. The approach adopted for testing the 3 boreholes at Altnabreac, which were drilled to 300m, is to measure hydraulic conductivity and pressure throughout the borehole. The zones of high pressure and hydraulic conductivity within the borehole were then selected for groundwater sampling.

The major ion and trace element chemistry of the groundwater is determined in order to judge the relative equilibrium of the groundwater sample and the rock. Stable isotope and uranium disequilibria measurements are used as corroboration or otherwise of the apparent timescales of water movement. The conclusions reached based on the hydrochemistry are then compared with the hydraulic results to see if a coherent picture emerges. The advantages of chemically derived conclusions are that they are more likely to represent relatively large volumes of rock and groundwater compared to the small volumes tested by single borehole hydraulic tests.

Certain drilling programmes proposed at future crystalline rock research sites include various borehole schemes designed for hydrogeological purposes. The concept is to distinguish between the several factors which contribute to the flow system by careful planning of borehole position and inclination so that the influence of individual parameters such as faulting, anisotropic fissuring, rock structure, lithological variations and topography can be determined independently. This should aid in future site selection since any site can be evaluated as showing varying interactions of a limited number of geological variables. At one of the proposed research sites a fully integrated meteorological, surface water, groundwater model will be constructed which incorporates geological information obtained during mapping. The objective is to identify nodes within the planned area which are most sensitive to the input parameters and position boreholes at or near these nodes. The results of hydrogeological testing and hydrological measurements should then have an increased sensitivity compared with more randomly placed boreholes.

At another research site the positions of an array of boreholes will allow inter-borehole studies which will be combined with fracture mapping and geophysics. This programme will be directed to the quantitative assessment of the influence of geological factors on hydrogeological parameters.

At the present time the work on crystalline rocks has enabled a basic methodology to be established and has led to the development of equipment to measure the parameters considered important. As yet the inter-relation between hydrogeology and geology is only qualitative but with the aid of surface and borehole geophysics and fracture analysis progress is being made towards quantitative links.

6. REPORTING PROCEDURE

The information so far obtained in the work is being co-ordinated in the form of a series of tasks each of which has an officer responsible for the preparation of a task report. At Altnabreac for instance there were about 20 separate tasks, reports on 11 of which have now been published or are nearing completion, while work is still continuing on the remainder. In line with government policy, IGS are issuing all the results for these tasks as they become available. The data and results are not formally published but are issued in a series of open-file reports which are available to those interested in the programme. They are distributed free of charge to anti-nuclear groups, and the general public, as well as to the scientific community, and altogether 16 reports were produced in 1980. They have been generally welcomed but have been critisized as being too technical. No attempt has been made so far to summarise the results in a non-technical form. No information meetings of the type held in the United States have yet been held but it is anticipated that, as the programme proceeds, both discussion meetings designed for the general public and non-technical review summaries of the work will be required.

7. CONCLUSIONS

The British programme is progressing very slowly as a result of delays inherent in the planning process. The strong local public opposition to the programme, based often on a mistaken view of its objectives has meant that elected representatives have only on one occasion approved a planning application to drill research boreholes. The resulting appeals are heard at local public inquiries and this can mean that even if permission is eventually given there are lead times of 5 or 6 years between a potential research area first being identified and the drilling of boreholes within that area. The delays are frustrating for scientific staff involved in the work but can be offset by staff involvement in supporting laboratory studies.

In 1980 opposition to the research programme and the involvement of geologists in it lead to widespread opposition to geological work generally and many other geological field programmes were adversely effected. It is hoped that the attempt by the Institute of Geological Sciences to make sure that people in areas most affected by the research work are better informed about the aims of the programme will at least result in a more objective debate during the coming months when geologists again go into the field.

At present the public debate in Britain is not whether or not the present programme is adequate to select and eventually characterise a site, but whether or not there should be a research programme at all. The most commonly held view among the opposition groups is that the safety of geological disposal can never be determined and that the waste should be stored on the surface for the next few hundred years and the present research programme abandoned altogether. The adequacy of the present programme has never really been debated or challenged. In Britain there is little debate on the information which needs to be obtained to select an eventual repository site or to demonstrate its safety. The present programme is designed to obtain sufficient information to warrant a decision on the feasibility of disposal in the context of British geology and to permit site selection and characterisation to be undertaken objectively at some time in the future if feasibility is proved. To talk of repository site selection in the British context is premature and it will be many years before a position is reached where the United Kingdom moves towards the selection of a site for high-level waste disposal.

REFERENCES

Gray, D.A. et al. 1976 "Disposal of highly-active solid radioactive wastes into geological formations - relevant geological criteria for the U.K." IGS Report No 76/12

Mather, J.D., Gray, D.A. and Greenwood, P.B. 1979 "Burying Britain's radioactive waste - the geological areas under investigation". Nature, Vol 281, pp 332-4

ACKNOWLEDGEMENTS

The work discussed in this paper forms part of the research programme into radioactive waste management sponsored by the Department of the Environment (acting on behalf of the Secretaries of State for the Environment, Scotland and Wales) and the Commission of the European Communities. It is published by permission of the Director of Institute of Geological Sciences (NERC).

DISCUSSION

F. GERA, Italy

In consideration of the difficulty in obtaining permission to carry out field work related to radioactive waste disposal in geological formation I wonder if some consideration has been given to changing the law that defines the procedure for obtaining drilling permits ?

J.D. MATHER, United Kingdom

When the Secretary of State for the Environment announced details of the current programme in July 1979 he stated quite categorically that all exploratory drilling would "be the subject of appropriate planning procedures". I can see no possibility of this very positive statement being changed whatever the difficulties involved.

L.J. ANDERSEN, Denmark

Do you intent to investigate all the sites before you will make a decision on a special site, do you have fund for such investigations, and how long do you think it will take before you finish ?

J.D. MATHER, United Kingdom

When the programme started it was the intention to investigate all the research sites mentioned in the paper before a decision on a possible demonstration repository site was made. This is still the intention but clearly the initial 10 years time scale is seriously behind schedule and it is not possible to predict how long the programme will now take to complete. The funds provided for the investigations are adequate but are subject to a serious annual underspend because of the difficulties is gaining access to research sites.

STOCKAGES SOUTERRAINS ET ENCOMBREMENT DU SOUS-SOL

Pour une gestion rationnelle de l'espace souterrain

Philippe MASURE

Chef du département Génie Géologique

au Bureau de Recherches Géologiques et Minières

RESUME

Comme pour tout autre aménagement, le concept de site souterrain favorable au confinement des radionucléides doit être défini précisément. Comme les meilleurs sites sont limités, tout spécialement les plus vastes, une gestion rationnelle de l'espace souterrain doit être mise en oeuvre afin de les préserver et les protéger.

SUMMARY

For every Kind of use, the concept of a suitable site has to be defined with precision. As the best sites are limited, especially the largest sites, a rational management of underground space should be implemented to preserve and protect them.

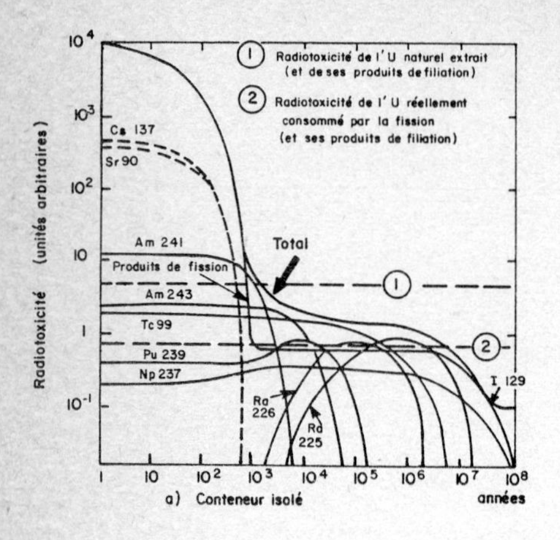

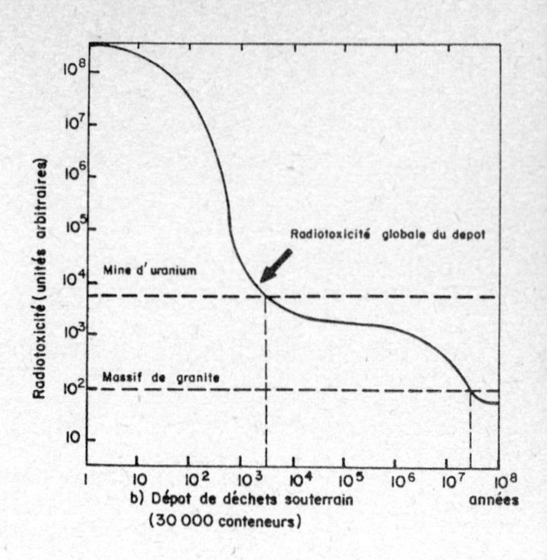

Fig. 1 - Evolution de la radiotoxicité des déchets de très haute activité en fonction du temps

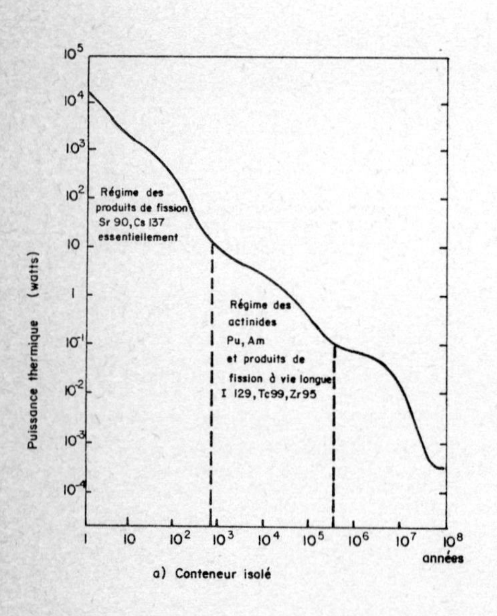

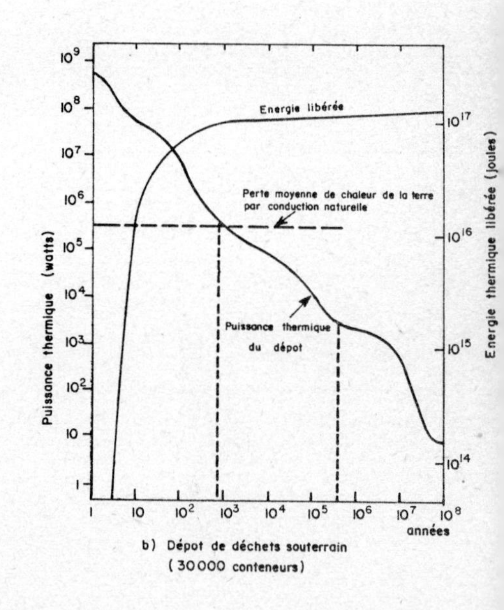

Fig. 2 - Evolution de la puissance thermique des déchets de très haute activité en fonction du temps

1 - INTRODUCTION

Les programmes de *stockage souterrain* de produits solides, liquides ou gazeux ont connu au cours de ces dernières années un développement remarquable, notamment en Europe. Que les raisons en soient technico-économiques, stratégiques, de sécurité ou de gestion rationnelle de l'espace, il est prévisible que l'exploitation des vides naturels (pores, fractures, karsts) ou construits (cavités minées) comme volumes de stockage dans le sous-sol va se poursuivre et même s'accélérer.

Les problèmes posés par l'évolution du milieu récepteur et son environnement au cours de l'exploitation du stockage (quelques générations humaines) ont donné lieu à des études scientifiques et techniques spécifiques qui ont permis de maîtriser avec une fiabilité suffisante l'impact de tels projets sur l'environnement. Seule la fracturation induite par certains stockages cryogéniques limite encore aujourd'hui le stockage souterrain de gaz naturel liquéfié à -162°C dans des massifs rocheux.

La situation est bien différente en matière d'*évacuation de déchets* de grande nocivité dans le sous-sol, leur confinement devant être quasi éternel. La production de certains déchets industriels nocifs a atteint un niveau tel* que leur dispersion dans le sol et les eaux n'est plus biologiquement satisfaisante. Leur transformation ou leur isolement définitif de la biosphère s'impose.

Coûtumiers et respecteux de normes de sûreté rigoureuses inhérentes à leur activité, les spécialistes de l'industrie nucléaire sont les premiers - et probablement les seuls jusqu'ici à avoir posé le problème des déchets de grande nocivité et à avoir initié les études scientifiques et techniques nécessaires.

Une partie des déchets solides produits au cours du cycle du combustible nucléaire renferme des *radionucléides à vie longue* (émetteur alpha) qui sont actuellement isolés de l'environnement par stockage en surface, dans l'attente de la mise en oeuvre de procédures d'élimination. La nocivité de ces déchets de faible ou moyenne activité persistera pour les générations futures durant des périodes qui s'étendent bien au-delà de l'avenir prévisible (plusieurs dizaines à plusieurs centaines de milliers d'années).

Les *déchets de retraitement* recèlent des actinides dont la très haute activité peut être considérée comme quasi-permanente. Ils contiennent également des produits de fission dont la décroissance se traduit par un dégagement d'énergie thermique considérable au cours des premiers siècles qui suivent leur production (fig. 1 et 2).

Parmi les concepts d'élimination variés qui ont été envisagés pour isoler ces déchets de la biosphère, l'enfouissement géologique profond est actuellement reconnu comme le plus accessible techniquement et le plus fiable scientifiquement, tout en paraissant économiquement raisonnable. De ce fait, une évaluation des formations géologiques potentiellement favorables au confinement des déchets a été faite dans la plupart des pays concernés mais rares sont ceux qui ont déjà sélectionné des sites d'évacuation définitive.

Les raisons de cette situation sont autant d'ordre scientifique que d'ordre politique. De nombreuses précisions et compléments

---

* La France produit annuellement 16 millions de tonnes de déchets industriels renfermant des déchets nocifs et 2 millions de tonnes de bains et concentrés chimiques qui présentent des dosages importants de substances toxiques et dangereuses

doivent être apportés à notre connaissance des sciences de la terre, des matériaux, de l'environnement et de l'homme avant de réunir les conditions d'un choix assuré de sites d'évacuation de déchets. Des efforts importants de recherche ont été faits dans ce sens pour les déchets de haute activité. Ils devront cependant être poursuivis durant une longue période, ce qui n'est pas incompatible avec leur cadence de production, relativement limitée.

Le problème est tout autre pour les déchets de faible et moyenne activité contaminés par des émetteurs alpha. Les quantités produites sont en effet très importantes et plusieurs centaines de milliers de mètres cubes devront être stockés à la fin du siècle dans la communauté européenne. Fort heureusement, ces déchets ne sont pas - ou très peu - émetteurs de chaleur et l'impact d'installations d'évacuation de tels produits sur le milieu géologique serait faible.

Il convient donc de porter en priorité les efforts de recherche sur le problème de l'évacuation géologique de ce type de déchets, et de considérer que la recherche et la sélection de sites de stockages de déchets de faible et moyenne activité constituent un problème urgent.

L'objet de cette note est double :

i - montrer que le choix des critères de sélection de sites d'évacuation des déchets radioactifs gagnerait à s'inspirer des exemples naturels de piégeage et confinement qui ont conduit à la concentration de substances minérales ou organiques dans la croûte terrestre.

ii - insister sur la nécessité d'une gestion rationnelle et planifiée des sites de stockage souterrain potentiels, l'encombrement croissant du sous-sol dans les pays de faible superficie pouvant rapidement faire passer de l'abondance à la pénurie des sites.

2 - QUELQUES REMARQUES SUR LES CRITERES DE CHOIX DES SITES D'EVACUATION

Les critères de sélection des sites d'évacuation de déchets radioactifs à vie longue en formations géologiques continentales ont été définies précisément en 1977 par un groupe d'experts réunis par l'AIEA [1]. Le document élaboré distingue des facteurs de sûreté géologique et hydrogéologique, des facteurs opérationnels et des facteurs socio-économiques. Pour sa part, la Commission des Communautés Européennes a réalisé, dans le cadre de ses programmes de recherche sur l'évacuation des déchets radioactifs en formations géologiques, un "catalogue européen des formations géologiques présentant des caractéristiques favorables" [2]. Ce catalogue a été élaboré sur la base de *critères de sélection* géologiques et hydrogéologiques et de *facteurs complémentaires de choix* liés à la géologie (occupation du sol et du sous-sol, risque d'intrusion accidentelle, sensibilité aux changements climatiques et hydrologiques).

La sélection s'est effectuée en deux phases :

i - choix du faciès (roches cristallines, formations argileuses sensu lato, formations salifères) ;

ii - sélection des formations favorables sur la base de critères éliminatoires (géométrie, homogénéité, sismicité, tectonique).

Les conditions hydrodynamiques et les caractéristiques physico-chimiques des formations sont des critères qui permettent d'affiner la sélection, mais dont on a rarement disposé de données précises dans cette phase d'analyse préliminaire. Il est pourtant certain que ces deux derniers critères sont parmi les plus restrictifs. Leur évaluation purement qualitative risque de conduire à une fausse impression d'abondance.

De grandes incertitudes persistent pour parfaire cette sélection. Les méthodes d'analyses, de mesures et de simulation mises en oeuvre au cours des dernières décennies pour préciser la connaissance des équilibres internes des massifs rocheux sont satisfaisantes dans le cadre de projets de grands ouvrages de génie civil, d'exploitation hydrogéologique ou de stockages souterrains dont la durée de fonctionnement est à l'échelle humaine. Il n'en va pas de même pour les projets d'évacuation de déchets radioactifs à vie longue pour lesquels la communauté scientifique se heurte à des incertitudes jusqu'ici méconnues :

- celles qui sont liées à l'*échelle de temps "géologique"* imposée par le problème, qui conduisent à tenir compte de situations, dimensions ou facteurs cinétiques mal connus et jugés négligeables dans les raisonnements courants : amplitude de l'érosion superficielle à l'échelle du massif par exemple, influence des cycles climatiques lents sur les équilibres internes du milieu, effet à long terme des variations de température et des contraintes sur les conditions d'écoulement des fluides en milieux de très faibles perméabilités, etc.

- celles qui sont liées au *dégagement thermique* des déchets de retraitement qui introduit une dimension nouvelle en matière d'impact des grands ouvrages sur la géosphère, que ce soit des points de vue structural, minéralogique, géochimique, hydraulique ou mécanique. Après la découverte du "choc hydraulique" induit par les retenues de certains grands barrages (Koyna par ex.) ou par l'injection massive de liquides dans certaines couches profondes (Denver, Colorado, par ex.), ne doit-on pas craindre un "choc thermique" méconnu jusqu'ici ?

Face à ces incertitudes inhérentes aux opérations nouvelles d'aménagement, la recherche de sites d'évacuation devrait tenir compte de considérations spécifiques :

i - la métallogénie enseigne que les concentrations minérales non syngénétiques (imprégnations, amas, remplissages filoniens) résultent du jeu de certains phénomènes géologiques dans des *structures de piégeage singulières*. La connaissance des potentialités de concentration et de piegeage naturel présentées par la géosphère (de nature structurale, stratigraphique, biochimique, capillaire, mécanique ou osmotique) a été à l'origine de l'option que représente l'évacuation des radionucléides à vie longue en formations géologiques. On doit toutefois remarquer que depuis lors, les références fournies par l'analyse détaillée des conditions naturelles de piégeage des éléments constitutifs de l'écorce terrestre ont été peu utilisées, aussi bien dans la conception des projets d'évacuation et de confinement des déchets que dans la définition de certains critères de sélection des sites potentiellement favorables. Les responsables du stockage souterrain d'hydrocarbures se sont pourtant inspirés avant nous des conditions de piegeage naturel (fig.3) pour rechercher des structures convenables. Le concept de *formations géologiques* potentiellement favorables au confinement des radionucléides (argiles, sel, granites, basaltes) devrait donc être complété par le concept de "*contextes géologiques*" favorables au piégeage des radionucléides [3].

Pour de nombreux aménagements de surface, le meilleur emplacement est plus ou moins *imposé par la nature* : défilés pour les barrages et les ponts, vallées et cols pour les voies de communication, chûtes d'eau pour l'énergie hydroélectrique. Tous ces sites naturels sont des emplacements privilégiés par des particularités morphologiques, tout comme les gisements métallifères sont privilégiés par leur contexte géologique et géochimique. Cette dernière notion gagnerait à être utilisée dans la recherche de sites d'évacuation, notamment pour les déchets de faible et moyenne activité : structures anticlinales ou litées étanches, zones à gradient hydraulique nul, barrières capillaires, filtres géochimiques, etc..

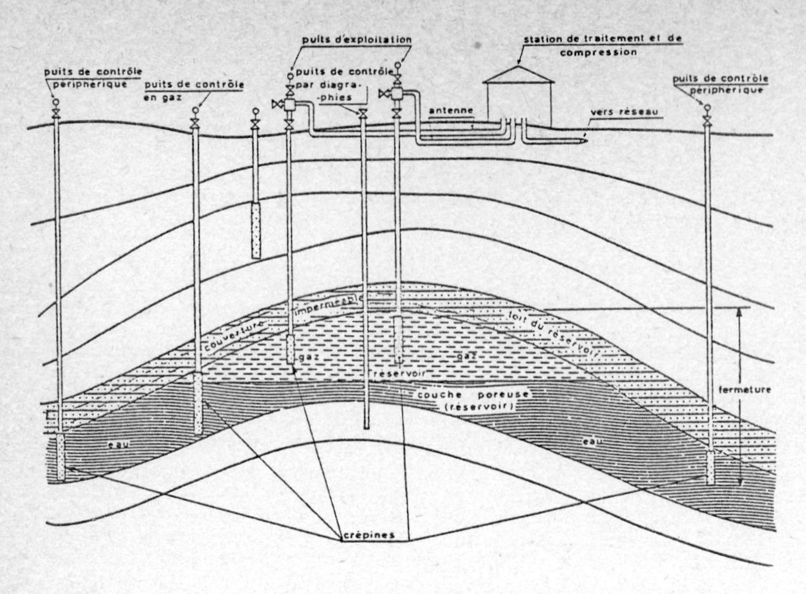

Fig. 3 - Coupe schématique d'un réservoir souterrain de gaz en nappe aquifère (document Gaz de France)

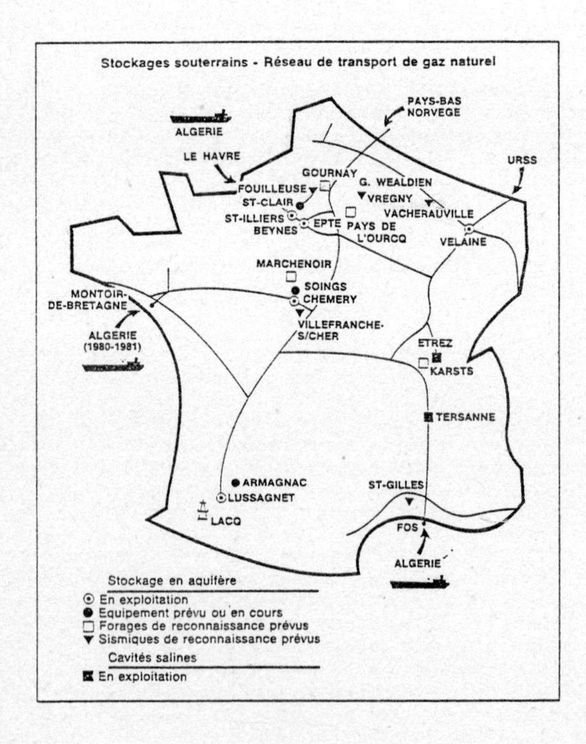

Fig. 4 - Installations de stockage souterrain d'hydrocarbures en France en 1980 (document Gaz de France)

ii - Le concept de barrières artificielles autour des conteneurs de déchets conditionnés n'a fait référence, jusqu'ici, qu'à des matériaux inclus dans les puits forés (barrières géochimiques ou métalliques. Il pourrait être enrichi, dans certains cas, par des solutions de *traitement du milieu géologique* entourant le dépôt en vue de limiter (ou éliminer) l'accès de l'eau souterraine dans le dépôt. L'aptitude de certaines formations à voir leurs caractéristiques naturelles améliorées par des procédés simples et dont la pérennité est assurée devrait constituer un autre critère de sélection préférentielle. A titre d'exemple, le traitement de certains milieux rocheux peut être très efficace par injection de silice ou de précipités dans les fissures ou les pores, par drainage des abords du dépôt ou par utilisation de barrières capillaires.

Certaines structures géologiques litées, judicieusement sélectionnées, pourraient également constituer une extension du concept multibarrière qui n'a pas été suffisamment évaluée et assurer ce que l'on pourrait appeler des *barrières géologiques multiples*.

iii - La définition de critères de sélection adaptés au problème de l'évacuation des *déchets de retraitement* ne sera possible que lorsque sera mieux connu l'impact thermique de dépôts de grande capacité. Si l'élévation de température dans le massif entourant le dépôt peut être diminuée par une limitation de la densité des conteneurs mis en place, l'énergie thermique dégagée en quelques décennies atteindra, dans le cadre de certains projets actuels de dépôts, une dimension comparable à celle de phénomènes géologiques non négligeables. L'expérience apportée par l'exploitation croissante des ressources géothermiques pourra sans doute améliorer les connaissances des spécialistes en ce domaine dans les années à venir. En attendant, on n'est pas en mesure d'affirmer que la probabilité est nulle de voir l'impact des dépôts de déchets de retraitement se traduire, dans certains cas particuliers, par des secousses sismiques dans les massifs rigides, par une forte migration de l'eau interstitielle dans les couches argileuses plastiques* pouvant conduire à une destruction de la structure interne de la formation, ou par des phénomènes de soulèvement néodiapiriques dans les dômes salins qui pourraient rompre l'étanchéité des couches imperméables qui ont empêché jusqu'ici la dissolution des masses salées par l'eau des nappes aquifères superficielles.

Face à ces incertitudes inhérentes aux opérations nouvelles en matière de génie géologique, il conviendrait de réaliser assez rapidement quelques *installations pilotes expérimentales et réversibles* simulant l'ouvrage final en vue d'analyser l'impact réel du dépôt sur la géosphère, avec auscultation détaillée du massif et de l'ouvrage pendant plusieurs décennies. Ce n'est qu'après une telle phase expérimentale que la définition de critères de sélection de sites d'évacuation de déchets de retraitement pourra être assurée avec pertinence.

3 - EXPLOITATION, OCCUPATION ET GESTION DU SOUS-SOL

En 1980, la France a consommé 108 millions de tonnes de pétrole et 24 milliards de mètres cubes de gaz naturel. Pour des raisons économiques et stratégiques évidentes, la France est actuellement contrainte de stocker 30 millions de mètres cubes de pétrole et 4 milliards de mètres cubes utiles de gaz naturel. Ce stock de gaz devra être doublé en 1985, triplé en 1990. Le recours classique à de grandes cuves de 50 m de diamètre et 20 m de hauteur nécessiterait un équipement de centaines de milliers de gazomètres. Leur emprise au sol serait de plusieurs centaines de km^2. En fait, 8 milliards de mètres cubes de gaz naturel sont déjà stockés en souterrain en France. Sur 18 sites de stockage, 7 stockages de gaz existent (fig.4) en formation aquifère et deux en cavités de dissolution dans le sel. Un stockage souterrain d'un demi milliard de mètres cubes de gaz représente seulement 30 hectares d'emprise au sol. Cet exemple suffit à démontrer que dans un pays de faible superficie comme la France, le recours au sous-sol ne constitue plus une simple alternative, comme il l'a long-

* qui ont une teneur en eau considérable : 20 à 35 %

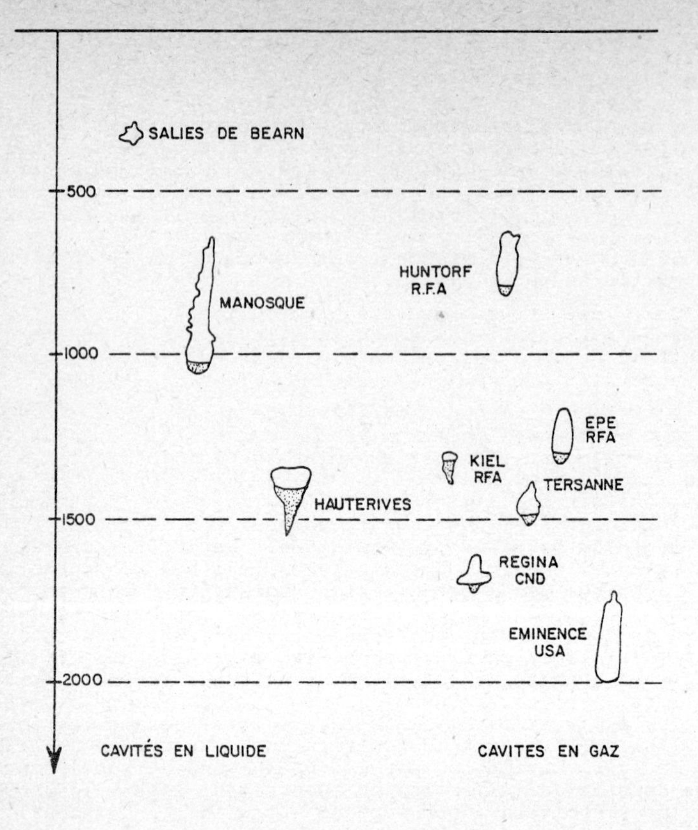

Fig. 5 - Forme, dimensions et profondeurs caractéristiques de quelques stockages d'hydrocarbures en cavités salines (d'après J. BERREST, Revue française de Géotechnique, 1981)

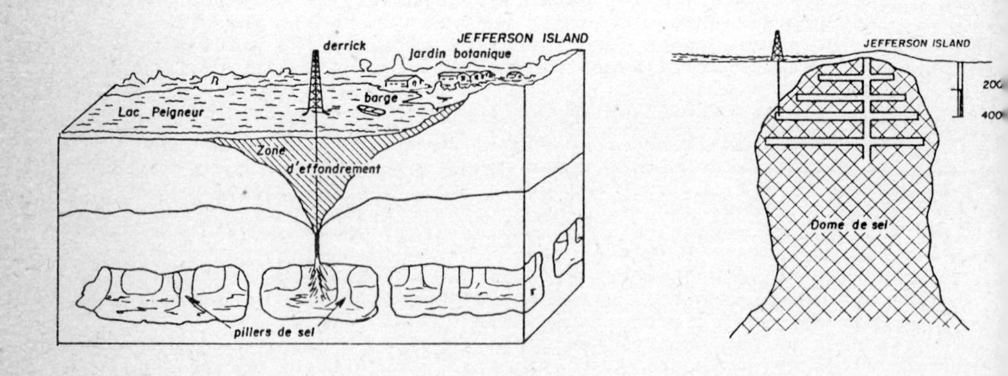

Fig. 6 - Exemple d'intrusion accidentelle dans une mine de sel entraînant la dissolution de piliers et l'effondrement de la mine (Jefferson Island, Louisiane, 20.11.1980)

temps été en matière d'aménagement. Bien au contraire, il devient aujourd'hui une nécessité impérieuse en matière de stockage de grandes quantités de produits consommables et de calories, tout comme il semble l'être en matière de confinement de déchets très nocifs.

Les premières traces d'exploitation du sous-sol remontent à plus de 6.000 ans où des mines de silex ont été découvertes en Belgique et en Hollande avec des puits verticaux de 10 à 15 m de profondeur. Depuis, l'exploitation minière s'est approfondie, avec des records à plus de 3.000 m (Afrique du Sud). Les réservoirs d'hydrocarbures sont recherchés jusqu'à 7.000 m de profondeur. Si l'exploitation des eaux souterraines conduit rarement à effectuer des forages de plus de 1.500 m, l'exploitation des ressources géothermiques, qui connaît une forte croissance, atteint fréquemment des profondeurs de 5.000 m. Le sondage de reconnaissances géologiques le plus long réalisé à ce jour est de 9.760 m (presqu'ile de Kola, URSS 1978), tandis que la mine à ciel ouvert (open-pit) la plus profonde a atteint -900m (Canada).

Les projets de stockage souterrain d'hydrocarbures (fig. 5) se situent généralement entre quelques dizaines (liquides) ou centaines de mètres (gaz) et 2.000 m (cavité de dissolution saline, Eminence, USA). Il est donc évident que les concepts actuels d'évacuation de déchets radioactifs se situent dans une tranche très superficielle de l'épiderme de la Terre*, quelque peu altérée par les activités humaines.

La profondeur des cavités minées est limitée par les problèmes de stabilité des excavations, par la chaleur naturelle dégagée en souterrain et par des considérations économiques. Des risques d'encombrement des sites de stockage souterrain en grandes cavités ne sont donc pas négligeables [4] dans un proche avenir et les massifs qui ne renferment pas de ressources minérales, organiques ou hydriques ne sont pas à l'abri d'une utilisation rapide de leurs *ressources spatiales* ou géothermiques. Une destruction irréversible des capacités de confinement de certains contextes favorables à l'évacuation est également à craindre.

Les exemples de coexistence difficile entre cavités souterraines sont déjà nombreux. Ainsi, la construction de tunnels hydroélectriques dans les Alpes françaises s'est heurtée à des difficultés liées à la proximité de nombreuses petites mines de charbon abandonnées depuis longtemps. Les effets catastrophiques de l'intrusion accidentelle d'un forage dans une mine ont récemment été illustrés de manière spectaculaire aux Etats Unis, en Louisiane. (fig.6). Le 20 novembre 1980, à Jefferson Island, les eaux du lac Peigneur disparaissaient dans un forage pétrolier en cours de perforation, détruisant par leur tourbillon plate forme de forage, derricks, barges, remorqueurs et maisons. Ce forage explorait les abords d'un dôme de sel, alors qu'une mine de sel exploitait le même diapir. La pénétration du forage dans la mine à 400 m de profondeur allait favoriser l'intrusion d'eau, la dissolution des piliers et l'effondrement partiel de la mine. Bien que les dégâts fussent considérables, il n'y eut pas de victime.

Toutes ces remarques démontrent l'importance qu'il y a d'organiser une *gestion rationnelle des ressources du sous-sol*. Celui-ci n'apporte pas seulement des ressources minérales, organiques et géothermiques (l'eau, les matériaux, les minerais, les hydrocarbures, le charbon), il apporte aussi des *ressources spatiales* et des *ressources de confinement*. Tout comme pour les ressources minières, incluant les eaux chaudes et l'eau, dont l'inventaire pour la France est réalisé par le Bureau de Recherches Géologiques et Minières, un véritable *inventaire des contextes géologiques* particulièrement favorables au piégeage et au confinement des radionucléides est à présent indispensable, les déchets nocifs devant être considérés comme un minerai artificiel. Cet inventaire devrait être intégré dans une banque de données centralisée regroupant tous les sites sélectionnés pour le stockage souterrain en grandes cavités convenant spécifiquement à chaque

---

* alors que l'épaisseur moyenne de la croûte terrestre est de 40 km

type de produit (hydrocarbures liquides ou gazeux, air comprimé, eau chaude, déchets chimiques, etc.) afin de définir des priorités et établir des réservations, des concessions. La gestion d'une banque de données sur les ouvrages et sites souterrains éliminerait les risques d'intrusion accidentelle et les difficultés de coexistence.

La prise en compte globale des intérêts de l'*exploitation* du sous-sol (ressources minérales, organiques et géothermiques) et de l'*occupation* du sous-sol (stockages, confinement, abris) permettrait d'en optimiser la gestion et d'éviter des conflits d'intérêt préjudiciables, par des mesures règlementaires, des incitations, des interdictions. Un véritable *bilan prospectif* en matière d'occupation de l'espace souterrain comparant les besoins et les ressources pourrait conduire à terme à une entente internationale qui éviterait de passer sans transition de l'abondance à la pénurie, situation que nous connaissons déjà en matière d'exploitation des ressources du sous-sol.

4 - CONCLUSION

Devant la production importante de déchets de faible et moyenne activité contaminés par des émetteurs alpha et devant l'accroissement net de l'occupation et de l'exploitation de l'espace souterrain, la sélection, le choix et la réservation de sites potentiels pour l'évacuation définitive des déchets radioactifs devient une nécessité urgente. Pour cela, des précisions et des compléments doivent être apportés aux facteurs de sélection et de choix qui ont été retenus pour faire les premiers inventaires. La notion de sites ou contextes géologiques de confinement doit en particulier être exploitée dans les concepts d'évacuation fixés par référence aux piégeages naturels de substances minérales, comme les gestionnaires des hydrocarbures l'ont fait pour leurs stockages souterrains en aquifères, en créant de véritables gisements artificiels. Si cet effort de recherche n'était pas réalisé dans les années à venir, on pourrait craindre un véritable divorce entre les soucis des scientifiques et des organes de sûreté d'une part et les impératifs des responsables de la gestion et de l'évacuation des déchets, d'autre part.

Une gestion rationnelle de l'espace souterrain et des ressources du sous-sol basée sur une évaluation prospective des besoins est fondamentale dans des pays de faible superficie comme les pays européens. On devra recourir à des plans d'occupation du sous-sol, afin de constituer de véritables réserves de sites souterrains. Faute d'une telle planification, des conflits d'utilisation des ressources du sous-sol et une véritable destruction de ses capacités de confinement sont à craindre dans un avenir qui n'est pas très lointain.

BIBLIOGRAPHIE

[1] Site Selection Factors for Repositories of Solid High-Level and Alpha-Bearing Wastes in Geological Formations", Technical Report Series n° 177, AIEA, Vienne 1977.

[2] Catalogue européen des formations géologiques présentant des caractéristiques favorables à l'évacuation des déchets radioactifs solidifiés de haute activité et/ou de longue vie", CCE, Bruxelles 1979.

[3] Ph. MASURE, P. VENET. - Considérations sur l'évacuation des déchets radioactifs en formations géologiques continentales. Première conférence européenne sur la gestion et le stockage des déchets radioactifs, CCE, 20-23 mai 1980, Luxembourg.

[4] P. DUFFAUT - Site Reservation Policies for Large Underground Openings, Underground Space, vol. 3, n° 4, pp. 187-193, Oxford 1979.

DISCUSSION

A. BARBREAU, France

Il y a personnellement une chose qui me gêne dans le principe de l'inventaire des ressources souterraines en vue de réserver certaines formations pour le stockage des déchets radioactifs. Il a été assez couramment admis dans la plupart des pays, qu'une fois qu'une opération d'évacuation serait réalisée dans une formation géologique, on n'assurerait le balisage et le gardiennage du dépôt que pendant une durée de temps limitée. On peut envisager de réserver des sites pour les déchets nocifs dans la mesure où l'on pourra conserver l'identification de ces sites aussi longtemps que durera la nocivité des déchets, mais on ne peut pas bâtir une politique sur cette base pour des déchets dont on admet la perte du balisage. Il semble que la politique de réserver des sites pour le stockage des déchets radioactifs soit contradictoire avec le principe même de la sûreté de ces stockages qui implique justement l'exclusion ou du moins la très faible probabilité d'intrusion sur ce dépôt.

F. GERA, Italy

Fortunately, the toxicity of radioactive waste goes down with time and therefore the requirement of isolation, which is absolute during the first few hundred years, is less stringent afterwards. The key of this particular problem is to perform a safety assessment and to determine what are the consequences if there is intrusion into the repository after different decay times. If we do this and we postulate intrusion after a period in excess of a few hundred years, we will find that the consequences are not really too serious and the overall risk might be considered acceptable. The decision of the acceptability of this risk is a political issue. The evolution of risk with time should be one of the elements that is important in deciding particular point.

My second comment deals with Mr. Masure's paper. He makes difficulties bigger than they are ; he says that there are many questions that have not been answered, yet, he claims that the possibility of isolating waste into geologic formations has not been demonstrated yet. I think that the capability of geologic environments to contain waste over very long periods is out of the question. This is demonstrated on a conceptual level by the many deposits of minerals of different nature that exist in the earth's crust. What we are dealing with is finding a specific site where we can put radioactive waste, so that we can be confident it can be isolated. What we now have to demonstrate is that a particular or specific site is suitable for that purpose.

G. de MARSILY, France

M. Gera et M. Masure ont parlé de l'existence de pièges naturels dans la croûte terrestre, qui permettent de penser que des déchets resteront confinés. Or chacun sait qu'une accumulation de mineraux ou de minerais dans le sol peut très bien se remettre à migrer. Ce sont des conditions particulières, qui à une date donnée, assurent l'existence d'un piège, mais l'évolution ultérieure de la croûte terrestre fait que ce piège peut être altéré. Alors comment ce concept de pièges peut-il être envisagé si l'on considère la possibilité d'une évolution de la capacité confinante des formations géologiques ?

P. MASURE, France

Je crois qu'il y a une réponse relativement simple à faire : ne retenir que des sites géologiquement stables, à l'échelle de la centaine de milliers d'années au moins.

F. GERA, Italy

I would like to add again that the time factor is an important issue and has an effect on isolation. If we consider the natural processes that might modify isolation, we know that in most cases it would require very long time periods. There are some exceptions and these will have to be covered within the safety assessment. We will have to estimate the probability, and consequences of time. If the geological barriers that still exist are as effective as they should be, if we consider their site acceptable, the migration time should be long enough to ensure that consequences are at an acceptable level.

J.P. OLIVIER, NEA

Pour faire un commentaire général sur ces questions, je dirais qu'il s'agit d'un problème qui a trait à la notion générale de risque. On se trouve en face d'une situation où on ne peut pas en tout état de cause aboutir à une sécurité absolue, et il convient donc en l'occurrence de faire une optimisation des systèmes telle qu'à la suite d'une analyse des risques aussi poussée que possible, on démontre que l'ensemble des mesures prises permet finalement d'aboutir à un niveau de risque qui est acceptable politiquement, je crois que c'est là la réponse à ce problème. On peut prendre un certain nombre de mesures dans le choix des sites pour trouver un emplacement qui a moins de chance qu'un autre d'être perturbé, soit du fait de l'action de l'homme, soit du fait de phénomènes géologiques à prévoir dans le temps. Tout est une question d'optimisation et d'analyse des diverses possibilités de perturbation du site pour essayer de déterminer quantitativement quel serait le dommage qui surviendrait à la suite d'une perturbation quelconque. C'est vraisemblablement dans cette direction qu'il faut s'engager, et je crois que les diverses réunions qui se sont déroulées précédemment, par exemple sur les scénarios de libération accidentelle, ont déjà permis d'avancer dans cette direction.

A. BARBREAU, France

Je vous suis tout à fait sur ce point et il est évident que c'est la seule politique logique. Ce qui m'inquiète ce sont les réactions des personnes qui disent : "vous ne pourrez jamais apporter la preuve que le stockage est sûr". Cela signifie qu'ils excluent effectivement ce concept de risque accepté. Ils voudraient une certitude absolue. On demande parfois aux spécialistes de faire la preuve que le stockage qu'ils envisagent de réaliser sera totalement sûr. Cela ne signifie rien, il n'y a pas d'activité humaine qui soit totalement sûre.

J.P. OLIVIER, NEA

Je crois qu'il appartient en fait non seulement aux experts, mais également aux autorités politiques, d'expliquer cette notion de risques qui finalement concerne chacun d'entre nous, dans toutes les circonstances de sa vie quotidienne. C'est une question d'éducation du public qui n'est sans doute pas spécifique au nucléaire. Il faudrait que les autorités gouvernementales, lorsqu'elles sont confrontées à de telles situations aient la possibilité de présenter les risques dans leur véritable contexte. Je crois que tout le problème est là : cela ne concerne pas seulement les personnes qui sont dans cette salle, cela concerne surtout les autorités qui sont chargées de prendre des décisions au plus haut niveau et qui doivent pour ce faire tenir compte de considérations multiples.

A. BARBREAU, France

Je crois qu'effectivement c'est un aspect tout à fait fondamental du problème dont nous discutons.

SESSION II

Chairman - Président

M.F. THURY

(Switzerland)

SEANCE II

SITING OF RADIOACTIVE WASTE REPOSITORIES IN THE IAEA UNDERGROUND DISPOSAL PROGRAMME

(Paper to be presented at the OECD/NEA workshop)

Z.Dlouhy, K.T.Thomas
International Atomic Energy Agency
Division of Nuclear Fuel Cycle

Abstract

The paper deals with the various activities of the IAEA in the field of underground disposal of radioactive wastes, their objectives and status of the different documents. Particularly, aspects of siting of repositories for high-level and alpha-bearing radioactive wastes in deep geological formations are described in more detail. Some areas, where further consideration might be useful, are briefly discussed.

1. Introduction

One of the most important activities related to the disposal of radioactive wastes is the selection of appropriate repository sites that will assure that disposal of wastes will result in no unacceptable detriment to man and his environment at any time. This means that a site for a deep geological repository should provide a high assurance for reliable prediction of a satisfactory long-term safety performance. It implies in general that the host formation in which the repository is located should be carefully selected and that the overall geological/hydrological system should be relatively well understood and amenable to a quantitative safety evaluation.

The overall objective of a siting process is the identification in a region of interest of one or more preferred host formations and sites which have a high probability of being suitable for the installation of an underground repository. The whole process should be conducted in a rational and systematic manner to ensure that the potential host rocks and sites are selected appropriately among those found in the region.

The siting process encompasses various activities ranging from planning and development of criteria through site investigations towards safety assessments including radionuclide migration studies. It should be pointed out that the most effective approach for site selection is to harmonize all activities relevant to the development of a repository. Thus throughout the process of site selection it is desirable that complete information on the type and quantities of wastes to be emplaced be known as well as the likely final form and packaging of wastes. Similarly, along with activities indicated above, the repository design should also be developed, starting from conceptual studies through a layout to the detailed design.

In general, the siting process is composed of various stages, starting with planning and general studies through area survey, preliminary site selection, and final confirmation. In each of these stages some site characteristics are considered which may lead to the rejection of unacceptable sites or areas and to the identification of the more suitable ones. The data required and complexity and sophistication increase as the site selection process advances towards its goal of identifying preferred sites.

Most of the areas indicated here are dealt with in various IAEA documents. An overview is presented in the following chapter. The Agency's underground disposal programme is based on the experience, plans and results from ongoing research and development work in its Member States as well as on the information from related programmes conducted by other international organizations, such as OECD/NEA, CEC, and CMEA. Many activities of the OECD/NEA in the field of geological disposal of radioactive wastes, as again evidenced by the holding of this workshop, provided valuable input to the IAEA programme, and continued co-operation in this field between both the organizations is important.

2. The IAEA Underground Disposal Programme

The IAEA from its inception has been actively engaged in the technical and safety bases of radioactive waste management including disposal of wastes. From a small beginning in the 1960's as more importance emerged to ultimate disposal technology, underground disposal has demanded major attention of the IAEA since the middle 1970's and is currently one of its major fields of activity.

In 1977 an integrated programme was initiated on the underground disposal of radioactive waste, ranging from the disposal in shallow ground and rock cavities to deep geological disposal. The objectives of the IAEA's work in this field are to develop a series of technical reports and guidelines, addressing under the disposal options five major areas (see Table 1):

(1) generic activities including regulatory activities, safety, assessments, basic guidance and criteria,
(2) investigation and selection of repository sites,

Table 1

IAEA UNDERGROUND DISPOSAL PROGRAMME

- Overview -

COMPONENT \ OPTION			SHALLOW GROUND	ROCK CAVERN	DEEP GEOLOGICAL FORMATIONS MINED REPOSITORY	LIQUID INJECTION	HYDRO-FRACTURING
Generic and Regulatory Activities, Safety Assessments ***	General	Basic Guidance	SR No. 54 (1981)				
		Regulation	*				
		Safety Assessment	SR No. (1981)				
		Basic Criteria	SR (1983)				
	for specific options	Guide to Disposal in	SR No 53(1981)	SR(1982)	**		
		Safety Analyses	TR (1982)	**	TR(1982)		
		Regulatory Procedures	**	**	SR No. 51	TR(1982)	TR(1982)
Site Selection and Investigation ***			TR No. (1981)	TR(1982)	TR No. 177 (1977) TR No. (1981) *	+	+
Design and Construction ***			SR (1983)	SR(1983)	*		
Operation, Shut-Down, Surveillance ***			SR (1983)	SR(1983)	*		
Waste Acceptance Criteria ***			*	*	*		

* Report(s) under consideration for 1983-84.

** No specific report is planned as subject may be covered in sufficient detail in other reports.

*** A safety standard document (Safety Series Category 1) is planned after 1984

\+ These reports will cover aspects of Site Investigation and Selection, Design, Construction, Operation and Safety Analyses.

Legend

SR Safety Series Report

TR Technical Report

SR No. / TR No. Safety Series or Technical Report Series already published or submitted for publication.

SR ( ) / TR ( ) Safety Series or Technical Report Series under preparation during 1981-82 with expected date of publication.

The possibility of combining reports in various subjects of contiguous areas has been and will continue to be considered for the topics mentioned, both horizontally (options) and vertically (components), taking into account the desirability of reducing the number of separate publications on related subjects and the timing for completion of particular subject reports. Symposia on broad subjects will be held and Technical Reports on specific topics prepared as required to collect background information.

(3) waste acceptance criteria,
(4) design and construction of repositories, and
(5) operation, shutdown, and surveillance of repositories.

It is supposed that the development of the above documents may cover a period of about 6 years, until 1984. In the second phase these documents will be re-examined in the light of experience gained in various Member States in order to arrive at appropriate Codes and Guides when the subject is sufficiently advanced and mature in all areas.

Regarding the disposal of radioactive wastes into deep, continental geological formations, several documents have been issued or are at present under preparation which address, or include consideration on, the siting of repositories. Most of them are Safety Series Reports, however, the IAEA is considering the preparation of a nubmer of supporting Technical Reports on various subjects of merit such as heat and radiation effects, pathway analyses , etc.

A Safety Series Report containing basic guidance on underground disposal [1] was prepared to provide general background for use by those authorities responsible for planning, approving and executing national radioactive waste management programmes, particularly for establishing radioactive waste repositories, and by those who must review applications and make recommendations regarding such repositories. Furthermore, this document may be useful to countries which are planning nuclear power programmes and which may be seeking general advice on the final disposal of radioactive wastes.

The first document within the IAEA programme devoted to repository siting was the Technical Report "Site Selection Factors For Repositories of Solid High-Level and Alpha-Bearing Wastes in Geological Formations" [2]. The main emphasis was dedicated to the different types of geological formations that might be considered suitable for disposal purposes and to the various factors that should be investigated to demonstrate the suitability of a formation.

Another report on the development of regulatory procedures for the disposal of wastes into deep geological repositories [3] was published in the IAEA Safety Series in 1980. This document was oriented primarily to guidance on what issues should be addressed in the licensing review and what guidance should be given to the applicant by the regulatory body in the course of the licensing procedure, including repository siting.

Mostly relevant to the subject of this workshop is a report dealing with site investigations for repositories in deep geological formations [4]. The report was completed in 1980 and is to be published in 1981. It covers the major activities, identification of information needs, brief description of the available scientific techniques for developing the required information, and methods for systematically applying these techniques to select, evaluate, and assess the suitability of specific sites. The investigations presented here are based on earth science; however, some other aspects regarding socioeconomic considerations are also briefly discussed.

Furthermore, a general document regarding safety assessment of underground disposal [5] will be published in late 1981. This document has a generic character and is devoted to various safety assessment approaches that can be useful at different phases of development of a disposal system and can help both the applicant and regulatory body in judging the appropriateness of the selected repository concept. A companion document [6] is under preparation on the principles and examples of safety analyses for repositories in continental geological formations. This document will present five examples, summarizing the analyses and the methods used, in order to broaden and enhance the understanding and application of these rapidly developing methodologies.

Another document under development is a Safety Series Report dealing with basic criteria for underground disposal. It will possibly contain an overview of different qualitative criteria to be applied at various steps of repository development.

In addition to these activities, a symposium on the underground disposal in 1979 in Finland was organized jointly by the IAEA and the OECD/NEA to collect new information and to review current development in this field[7]. From most of the papers it could be stated that an impressive investigative work was carried out and that a broad scientific base exists at present behind the underground waste disposal. A technical document on regulatory aspects [8] was also issued in 1980 to present information on national regulations and approaches including the role and responsibilities of various organizations involved in licensing, inspection and other related problems.

3. Some Considerations Regarding the Repository Siting

An idealized sequence of activities for underground repository siting is presented in Table 2. Hence it can be seen that the IAEA with the assistance of experts from Member States is covering most fields by specific documents as shown in Chapter 2. Several observations regarding some of the activities shown are discussed below.

Planning and General Approaches

A plan for siting of underground repositories developed at the beginning of the process usually includes

- identification and description of the tasks to be performed on the project,
- sequence diagrams showing relationship between the various tasks (e.g. site characteristics to be considered at each phase of the process),
- criteria adopted for the area survey, preliminary site selection , and site confirmation listed for each site characteristics,
- outline of the procedures for applying these criteria and factors and sources of information needed for their application,
- comprehensive time schedule.

Selection of appropriate sites for underground repositories involves integration of programmes of site investigative work in many disciplines. These include branches such as earth sciences, engineering, safety analysis, health physics, ecology and social sciences. The investigation encompasses theoretical, laboratory and field activities carried out in a general stepwise fashion but with significant interaction between steps.

From some possible ways of selecting potential sites and for implementing various investigation studies, a stepwise approach may be chosen as an idealized one. This approach is intended to serve as a general guide for developing national programmes. It does not exclude from consideration by national authorities other possible approaches through which the location of preferred sites may be reached without going through a highly systematic sequence of investigations starting with studies of large areas.

In larger countries with complex geology there may be a need to refine the overall siting process in several steps. In smaller countries or in countries with extensive geological and hydrological information, it may be possible to identify specific potential sites directly. It may further be possible to narrow the location process very quickly in some countries wherein non-geological factors may dominate the overall selection process.

A comparison of various approaches may be very useful (e.g. as a complementary document to the report on basic guidance for underground disposal [1]).

Development of Criteria

The IAEA Secretariat in consultation with the Technical Review Committee on Underground Disposal felt the urgency of laying down criteria on various aspects of disposal of radioactive wastes. Therefore, the process of trying to develop these criteria was started, and is planned to be completed in 1982.

Table 2 Idealized Sequence of Activities for Repository Siting

STEP	Planning and General Studies	Area Survey	Preliminary Site Selection	Site Confirmation
OBJECTIVE / RELATED ACTIVITY	Develop overall plans and criteria and review basic data	Select potential repository areas	Select potential sites	Confirm acceptability of final site/s/
PLANNING	Establish plans and timetable	Modify and expand plans as necessary	Modify and expand plans as necessary	
CRITERIA DEVELOPMENT	Establish generic waste acceptance and repository performance criteria, establish criteria for area survey	Establish criteria for preliminary site selection	Establish criteria for site confirmation	
WASTE FORM DEFINITION	Define waste immobilization methods	Define waste conditioning methods	Select final waste conditioning	Design waste packaging
REPOSITORY DESIGN DEVELOPMENT	Develop repository concept	Develop conceptual repository designs for potential host rock types	Develop preliminary repository designs for potential sites	Develop detailed repository design for final site/s/
SITE INVESTIGATION	Review site investigation techniques	Select site investigation techniques appropriate for potential host rock types	Expand site investigation techniques as required	Expand site investigation techniques as required for final site/s/
SITE SELECTION	Review site selection factors and identify potential host rock types	Select and inventory potential host rock types	Select potential sites	Select final site/s/
		characterize potential areas	characterize and compare potential sites	characterize in detail final site/s/
RADIONUCLIDE MIGRATION	Review nuclide migration models	Modify nuclide migration models for potential host rock types	Apply models for potential sites, test models with data	Apply model for final site/s/
SAFETY ASSESSMENT	Review/select safety analysis methods	Perform generic safety analysis for host rocks types	Perform preliminary safety analysis for potential sites	Perform detailed safety analysis for final assessment of site/s/
REGULATORY ACTIVITY	Provide basic regulation	Review information and data provided by the implementary organization	Issue preliminary decision upon suitability of potential sites	Review in detail the safety report and decide upon suitability of the final site/s/

All criteria presented in a draft document prepared by the IAEA Secretariat are of a qualitative character and it is for national authorities to assign to them quantitative values depending on specific site conditions and waste characteristics, taking into account that this quantification shall be based upon the overriding radiation protection principles.

Selection of Potential Sites

In site selection, two approaches may be followed for collecting the information, "parallel approach" and "serial approach". In the parallel approach all the necessary information is collected for all areas or sites. In the serial approach the information is collected only for areas and sites not rejected previously. The advantage of the parallel approach is that it is not necessary to await the result of previous rejection for collecting additional information or information on other site characteristics; the advantage of the serial approach is that the amount of work involved is smaller than with the parallel approach. In actuality, work would usually move from the parallel approach to the serial one.

The screening of potential sites may be accomplished by an intercomparison of the sites using elementary techniques of suitability scaling and simple comparison methodologies. The employment of such methodologies is particularly useful in narrowing the list of potential sites to a few candidate sites. It should be noted, however, that there is no universally accepted methodology for conducting this comparison. Each method has advantages and disadvantages.

In order to enable appropriate authorities to make a judgement upon the suitability and acceptability of a site a guidance on various screening and comparison methodologies may be useful in the future.

Safety Assessments

Final assessments of the safety of repository sites are based on detailed safety analyses. They are applied in various forms throughout the overall process and their purposes are to estimate potential radiation exposures to man for the present as well as for distant future, to compare these results with basic radiation protection criteria, and to present the results for judgement by an appropriate regulatory body.

It may be pointed out that valuable input to the safety analyses may be obtained not only from a repository site, but also from similarly selected research and reference sites. Direct field investigations and experiments, for instance regarding radionuclide transport, may be required both in development of confinement and repository concepts, and in producing experimental evidence to check and verify model calculations when repository sites are selected or developed.

Regulatory Activities

The national authorities have to establish the legislative basis for the fundamental decisions supporting the national disposal concept. They have to assign various responsibilities among the waste producers, operators of the repository, and regulatory bodies. Independence of the regulatory bodies from the others should be ensured. At relevant stages of the investigations, appropriate regulatory bodies should be kept informed and involved in decisions.

Appropriate authorities are provided with sufficient information to permit decisions on authorization for design and construction of a repository. The procedures which could logically be followed in reaching a set of rational decision by a regulatory body are generally dealt with in the appropriate IAEA Safety Series Report [3], however, a more detailed guidance on regulation is necessary and will be prepared in the next future.

4. References

[1] Underground Disposal of Radioactive Waste: Basic Guidance
IAEA Safety Series Report No.54(1981)

[2] Site Selection Factors for Repositories of Solid High-Level and Alpha-Bearing Wastes in Geological Formations
IAEA Technical Reports Series No.177(1977)

[3] Development of Regulatory Procedures for the Disposal of Solid Radioactive Waste in Deep, Continental Formations
IAEA Safety Series Report No.51(1980)

[4] Site Investigations for Repositories for Solid Radioactive Waste in Deep, Continental Geological Formations
IAEA Technical Reports Series (in press 1981)

[5] Safety Assessment of the Underground Disposal of Radioactive Wastes
IAEA Safety Series Report (in press 1981)

[6] Application of Safety Analyses for Radioactive Waste Repositories in Continental Geological Formations
IAEA Technical Reports Series (in press 1981)

[7] Underground Disposal of Radioactive Wastes
IAEA/NEA Symp.Proc., Otaniemi, IAEA Vienna 1980

[8] Regulatory Aspects of Underground Disposal of Radioactive Waste
IAEA-TEC-DOC-230 (1980)

OECD/NEA Literature Relevant to the Subject:

- In Situ Heating Experiments in Geological Formations
  Proc. of the Ludvika Seminar,Sweden, 1978
- Migration of Long-Lived Radionuclides in the Geosphere
  Proc. of the Brussels Workshop, Belgium, 1979
- Low-Flow, Low-Permeability Measurements in Largely Impermeable Rocks
  Proc. of the Paris Workshop, France, 1979
- Borehole and Shaft Plugging
  Proc. of the Columbus Workshop, USA, 1980
- Radionuclide Release Scenarios for Geological Repositories
  Proc. of the Paris Workshop, France, 1980.

SITE INVESTIGATIONS FOR A NUCLEAR WASTE REPOSITORY IN CRYSTALLINE ROCK

O. Brotzen
Consulting Geologist
Djursholm, Sweden

ABSTRACT

Information and documentation needed to describe a given site is identified. A tentative sequence of investigations may comprise the following steps:

Preliminary testing of the site by a few geological and geophysical traverses and a deep drillhole. If encouraging, detailed surface geological and geophysical characterization of the site is followed by shallow drilling directed at suspected fracture zones in the bedrock. The availability and nature of sound rock at depth, and the character of selected fracture zones there, is established by a limited number of fully cored drillholes and appropriate geological, geophysical, hydrogeological and geochemical studies. The site is then judged by an overall safety analysis rather than by a fixed set of predetermined criteria. A concluding step of selected verification studies is also suggested.

1. Introduction

The nature of a site, the type of repository considered and the locally available facilities will generally decide what investigations will be made at a specific site. The present paper tries to outline the purpose, scope and sequence of such investigations, and a selection of methods, which have been employed in Sweden in work considering high level waste and spent nuclear fuel. It, therefore, reflects natural conditions often met with in this country, such as moderate topography, a relatively uniform crystalline bedrock, composed mainly of granites and gneisses, insignificant weathering, and a soil cover, which normally is less than 15 m in thickness. Groundwater conditions are characterized by a free groundwater table, often found within a few metres beneath the surface. Even within this general framework, site-specific conditions may call for substantial changes in the general approach. The following outline, therefore, may serve primarily as a basis for more detailed planning, and also for a more general discussion on site investigations in similar environments.

2. Technical background

The plan of a repository will consider the expected nature and quantity of the waste, and the geological conditions at the suggested site. Such planning is facilitated by a modular design, which permits the final arrangement of the modular elements to be adapted to the geological structures as actually found in later underground exploration. For the present purpose a technical presentation of the design and dimensions of these modular elements, their general mode of combination and operation, and their overall requirements for underground space, will be sufficient.

The depth of the repository is essentially decided by safety considerations, primarily regarding the groundwater flow and a suspected transport of radionuclides to the surface by the groundwater. They will consider both the natural conditions and the perturbations by the repository and the waste. The analysis of these conditions is the subject of later sections. The depth, therefore, may be finally decided upon after the local conditions have been analysed, such that overall safety requirements can be fulfilled.

Regarding the impact of the repository, its mechanical and channelling effects are practically eliminated by the backfill developed for this purpose. Its characteristics and technical application are briefly described on the basis of existing information.

The heat effects of the waste are kept low by the combination of a prolonged period (40 years) of supervised near-surface storage prior to emplacement, and the spacing of the waste in the repository, leading to a rather low heat load, thereby keeping the temperature in the repository below 100°C.

Calculation of the thermal impact of the repository requires data on the thermal conductivity, capacity and expansion of the host rock. They can be based on values from literature, but their local validity may be confirmed by measurements on samples from the site.

The rock-mechanical aspects of the repository and its design likewise can be estimated on the basis of values of the mechanical properties of the host rock as found in literature. Again, verification by actual data from the site is desirable. In addition, the local natural stress-situation in the rock and its variation with depth may be investigated.

Documentation:

Review of the nature and quantity of waste to be handled at the repository, including canister materials and design.

Technical, constructional and operational information on the repository and its overall dimensions.

Technical information on the application and properties of backfill and buffer.

Report on thermal effects of the repository with review of basic data.

Report on rock-mechanical effects of the repository, including basic data on the host rock and the role of the natural stress-situation.

3. Safety-determining factors

The safety analysis for a given repository at a specific site comprises a complex set of calculations. They serve to quantify the environmental impact of the repository and to evaluate it by comparison with recommended standards of radiation protection and local and regional levels of natural radiation. It may be made to provide a limiting set of parameters, which a repository at a given site has to meet in order to be declared acceptable. The decisive site-specific factors in this respect are:

- The flow volume of groundwater that may contact the waste and carry its radionuclides to the surface.

- The concentration of each nuclide, which the water may carry, given the prevailing chemical conditions and the chemical reactions with the host rock;

- The lifetime of the canisters;

- The time of transport for each nuclide to reach the surface.

The last point decides the extent of the en-route radioactive decay. The transport time itself is the product of the flow-time of the groundwater and the retardation factor of the nuclide, reflecting its en-route sorption. Complications in these calculations for some of the nuclides arise from their radioactive mother-daughter relationship.

The natural dilution of the groundwater from the repository by other waters en-route and at the surface is an additional site-specific factor in the safety analysis.

The flow volume and flow time of the groundwater, as well as its dilution, reflect the geometry and hydrogeology of the site. The expected nuclide concentrations, the lifetime of the canisters and the retardation factors reflect its geochemical conditions. Therefore, hydrogeology and geochemistry represent essential fields of site investigation.

Both hydrogeological and hydrochemical factors are susceptible to improvement by various engineering measures. For instance, chemical reagents may be added to the buffer and backfill materials to improve the retention of specific nuclides. This type of chemical engineering, however, is a part of the containment concept rather than of the site investigations. On the other hand, the flow volume of groundwater may often be markedly decreased, and the flow time to the surface increased, by increasing the depth of the planned repository at a given site. Hydrogeological modelling, cf. section 4.2.4, may, therefore, consider alternative values of repository depth.

Documentation:

A list of limiting values of decisive factors or their products, which the site is claimed to meet.

4. Site-specific data

4.1 General

Basic data on a given site would include a brief description of the location of the site, indicating neighbouring settlements and townships, land ownership and use, available roads and other means of transport, electric power and so on, as well as a characterization of the topography.

The required information can probably be compiled from generally available sources. Then no actual investigations will be needed. The description should be accompanied by a standard topographical map in a scale of about 1:50 000 of the site and the surrounding region as well as a special map of the site and its closer surroundings in a scale of about 1:2 000-5 000 and with contour-lines at one metre equidistance in flat country,as a basis for later work. This map, reduced to 1:10 000,would also serve as a base map for later detailed geological and hydrogeological presentations. Already the basic version may show, in addition to topographic details, features of special interest such as roads and pathways, ownership boundaries, walls and fences, houses and other permanent and semi-permanent installations, natural springs and water wells, existing excavations, boreholes and natural outcrops of the bedrock. The detailed base map may, therefore, require special mapping, aerial photographs, some levelling and certain field checks.

Documentation:

Brief general site description.

Topographical map, approx. 1:50 000.

Detailed base map 1:2 000 1:5 000, plus reduced version at 1:10 000.

4.2 Geology

4.2.1 Regional setting

Data on the regional setting of the site regarding geomorphology, climate and surface hydrology, recent changes of the level of the land, seismicity, nature and distribution of superficial deposits, known mineral resources, main rock types and main structural features of the bedrock, as well as population density and transportation network, may all be displayed on a series of maps at scales of 1:10 000 000 to 1:250 000. The maps should be accompanied by brief explanatory notes, focussing on features with a bearing on the suggested repository and the future development of the site.

Most of this information can probably also be represented by or compiled from existing sources with minor supplementary work, e.g. structural interpretation of satellite imagery.

Documentation:

A series of small-scale maps with explanatory notes.

4.2.2 Local geology

The local geology at the site and its nearest surroundings may be shown on a standard geological map of about 1:50 000. This should provide an overview of the nature and distribution of the superficial deposits in the area, and also outline the available outcrops of bedrock and their geological designation as well as the main faults and fracture-zones, excavations and locations of mineral and water production, which can be represented on this scale.

In some cases, a separate structural map in the same scale, showing major structural features, such as lineaments, faults, fold trends, bedding,foliation, etc. may be called for. Also, a few schematic profiles showing the expected configuration of the candidate host rock and its flanking formations at depth may be prepared to accompany the standard geologic map. A separate map of about 1:10 000 is required to show the bedrock geology only, displaying the inferred extent and structure both of the candidate host rock and of the surrounding formation, as if stripped of their soil cover. Special attention should be given to the location and extent of observed and suspected main faults and fracture zones of the area.

The standard geological map in many cases will already be available in published form, whereas the 1:10 000 map may require actual field work and detailed analysis of aerial photography. The main rock-units, including the candidate host rock, should be briefly characterized at this stage, regarding general nature, and gross structural, hydraulic and compositional features.

Documentation:

Standard geological map, 1:50 000, tentative vertical sections (profiles) show overall expected bedrock configuration at depth.

Detailed map 1:10 000 showing bedrock distribution and main structures as well as hydrological features, at the site and its surroundings.

Explanatory notes with general description of important rock-units, including mineral resources.

4.2.3 Host rock conditions

The features of the host rock at the site and its nearest surroundings comprise both those of the rock itself and of its main discontinuities, such as faults and fracture zones. The discontinuities are generally considered to be of special importance in regard of the geotechnical and hydrogeological conditions at the site.

Their location, extent and direction are, therefore, investigated by geological and geophysical surface mapping at 1:2 000. Their continuation in depth is traced by drilling, shallow at first and deeper later. This work is supplemented by observation in outcrops regarding the nature, extent, spacing and direction of small-scale fracturing in the rock.

The drill cores also provide ample material to study the mineral and chemical composition of the rock, as well as its compositional and textural variation. Also, the nature and depth of weathering is investigated in the cores, as well as in the outcrop, and so is the occurrence and nature of rock alterations and fissure fillings.

Documentation:

Map, 1:5 000-10 000, of host rock variation and main discontinuities at surface, with location, inclination and horizontal projection of existing drillholes, possibly also with horizontal projection and indication of the depth of their intersection with main discontinuities. Supporting geophysical maps and profiles.

Vertical sections, 1:2 000 to 1:5 000 through main discontinuities and relevant drillholes.

Geological and geophysical logs of drillholes and cores, 1:50 to 1:2 000, with data on fracture frequency, inclination (orientation), mineralization and host rock variation and heterogeneities.

Diagrams on orientation of small fractures, with notes on spacing, length, morphology, filling and alteration.

Description of host rock and discontinuities with mineral identification (thin section, X-ray, DTA differential thermal analysis) and chemical data.

4.2.4 Hydrogeology

Hydrological background information of some interest in regard of the local groundwater situation includes locally valid data on the annual precipitation and evapotranspiration, as well as measured run-off data pertaining to that part of the drainage system sustained by the site and its surroundings.

More important is the configuration of the free groundwater-table together with the variation of the density and temperature of the groundwater with depth. They are required for calculation of the hydraulic potential throughout the rock volumes at the site. The hydraulic potential combined with information on the hydraulic conductivity of the host rock and its discontinuities defines the volume of the groundwater flow through the different parts of the host rock and its waterbearing zones. With reasonable values of the flow porosity also the velocity of the water, and hence the flow-time from a repository at depth to the surface may be estimated.

The actual calculations, disregarding rough order-of-magnitude estimates, require extensive computer modelling. The results may be verified by independent, rather time consuming, field determinations of the hydraulic potential, flow volumes and flow times and additional measurement of the hydraulic conductivity in selected parts of the groundwater system. Modelling of the groundwater flow should comprise both the undisturbed natural conditions and the perturbation brought about by the heat production of the emplaced waste. The calculations should consider a number of alternative depths for the repository. If the isolation of the waste, e.g. by a sufficient lifetime of canisters, will outlast the thermal impact of the repository, thermally induced groundwater flow need not be modelled.

Documentation:

Maps and sections, 1:2 000-10 000 representing the hydraulic potential and the groundwater flow (pattern, volume and time distribution in plan and depth, as derived from computer modelling of the site) with explanatory notes.

Map of groundwater-table, indicating observation points, water wells and natural springs, areas of recharge and discharge and the probable location of the groundwater-divide, i.e. the boundary conditions of the flow system.

Report on the hydraulic conductivity, temperature and density variation, as measured in available drillholes, with supporting geophysical logs.

Report on verifying local measurements of hydraulic potential, flow volume, flow velocity and flow porosity.

Discussion of what changes in the hydrogeological system would arise from geologic and climatic events that may occur during the lifetime of the repository.

4.2.5 Geochemistry

Data are needed to quantify the chemical interactions between host rock, groundwater, buffer substance, canister materials and waste. Apart from the chemical characterization of the host rock, cf. section 4.2.3, the site-specific data relates primarily to the chemistry of the groundwater and data describing the distribution of dissolved nuclides between the water and the rock.

The spatial variation of fundamental parameters in the groundwater, such as temperature, pH, Eh (redox potential) and electrical conductivity (reflecting solute concentration, ionic strength and density) is established by appropriate logging of the boreholes. Also, the natural radioactivity of the rock and the radon content of the water may be determined in this way.

The data are supplemented by laboratory data on samples of the water at depth, stating the content of major and minor constituents, including sulfide, gases, organic material and microorganisms. Age determinations of such samples by isotope methods are also of value as independent information on the groundwater movements at the site.

Much information has been collected in the last few years on the interaction of dissolved nuclides with various host rocks and individual minerals. The local validity of the data may be confirmed by supporting laboratory study and field experiments.

Documentation:

Report on the chemical conditions in the groundwater system at the site and its surroundings based on borehole data and laboratory determinations, including a discussion on the expected performance of buffer substances and canisters in this environment under repository conditions.

Report on expected interaction between dissolved radionuclides and the host rock based on available data, with recommended input to the safety analysis.

Discussion of what changes in the geochemical system may arise from geologic and climatic events during the lifetime of the repository.

5. Investigations

5.1 Scope

The following overview considers the investigations to be made from the surface at a site to establish that suitable conditions exist for a waste repository. It considers a repository in igneous and metamorphic rocks, for heat-producing, long lived high level nuclear waste. It does not consider legislative, societal or economic investigations regarding the site. Likewise, it does not consider the site selection process, nor the basic compilation work involved in the general description of the site and its regional geologic setting. Finally, the laboratory and office work required to support and evaluate the site investigations will not be covered here.

5.2 Surface investigations

The surface mapping comprises production of a detailed base map at a scale of 1:2 000 to 1:5 000, as well as a detailed geological mapping, including photo-geological study, and surface geophysical investigations, cf. sections 4.2.2 and 4.2.3. Only the geophysical work needs further comment here.

It serves to detect and map waterbearing fracture-zones in the host rock, which are concealed beneath the surface. An electrical resistivity survey will pick up such zones by their anomalous conductivity. Induced polarization work or seismic profiles across such conductors in the bedrock will further show if they are indeed caused by waterbearing fractures or if they may be due to other causes, such as diseminations of conducting minerals. Seismic profiling may also be used to indicate depth to bedrock. Electromagnetic profiles across the conductors will indicate their inclination.

A magnetic survey over the host rock will show significant structural trends in the host rock.

Within a few years, radar equipment is expected to be available to map water-bearing fractures and to determine depth to the water table and to bedrock. When proven, it may replace some of the methods indicated above.

At present, a magnetic and resistivity survey covering the site with ample margin is generally recommended. Induced polarization is applied with the same electrode settings, and, therefore, may advantageously be made to show the distribution of polarizable minerals at the same time as the resistivity work. When the results of these surveys are available, a layout for seismic and electromagnetic profiling is decided upon jointly by the geologist and geophysist in charge. The whole planning of the geophysical program must consider the geologic conditions at the site, notably concerning the nature of the bedrock, the depth of weathering and the depth of the soil and other superficial deposits.

The results of the surface investigations are represented in detailed maps (1:10 000) showing superficial deposits, hydrological features, bedrock distribution, main structures and discontinuities, with expected inclinations and suggested location and inclination for shallow drillholes. Supporting geophysical maps and profiles and diagrams on the fracture pattern in the host rock are also produced.

5.3 Shallow exploration

The shallow exploration basically relies on boreholes (about 10 cm diam.) to about 100 m depth or somewhat deeper, that can be drilled with down-the-hole rotary air hammer rigs, which also can drill inclined holes. Such holes are located at hydrogeologically decisive points, and along known and suspected fracture zones, in order to verify their presence and determine their course, inclination and hydrogeologic character.

In drilling the following information should be collected:

- Exact location and inclination of the hole (local grid system).
- Record of drilling progress.
- Samples of well cuttings at regular intervals (5 m) and when changes in their nature or colour occur.
- Depths at which water is struck.
- Static water level at the beginning and end of each drilling period.
- Final depth of hole.

After completion of a hole, it is flushed repeatedly with compressed air from near the bottom of the hole, whereupon the water is allowed to recover. The progress of recovery is recorded. The water level is monitored. After a period of stabilization, each hole is logged for caliper, temperature, electrical conductivity of the water, pH, Eh and accoustic velocity, point resistance and spontaneous polarization in its walls. Other logs may be desirable and made at the same time.

The results of the logging will guide the testing of the hole, and selected sections, by water injection tests.

After a prolonged period of stabilization, pumping of some of the holes and of selected sections is made to extract samples of the water, and to study the hydraulic drawdown as well as the response in the surroundings, surrounding holes being used as observation wells. The samples, both of the water and of the well-cuttings, are examined in the laboratory. The total information obtained is evaluated regarding the exact location, course and direction of existing waterbearing zones, their hydrogeologic character and role, and the geochemical situation at the site.

5.4 Exploration at depth

Exploration at depth at this stage relies entirely upon a limited number of deeper boreholes, drilled to and well beyond the depth intended for the repository. The small number of boreholes is dictated by the desire to limit perforation of the host rock.

The boreholes (diam. 56-76 mm) are normally made by diamond drilling, with maximum core-recovery by the use of double or triple core barrels. The water used for flushing during drilling should be analysed and marked by a suitable tracer substance to aid in later evaluation of hydrochemical data.

The location and direction of the boreholes is selected to verify the presence at depth of technically attractive volumes of sound rock of low hydraulic conductivity, and also to locate some of the major fracture zones detected near the surface. Deep drilling, therefore, logically comes after the completion and evaluation of the shallow exploration.

It may often be advantageous, however, to start the drilling of a first deep hole already when the results of the first geological and geophysical traverses have been evaluated. This is done in order to verify at an early stage the general potential of the site. If such an early borehole proves encouraging, drilling may be continued down to the maximum depth to be investigated. If, instead, the hole proves discouraging, the complete program may have to be discontinued and the site abandoned. A new site will then be needed.

In drilling the deep holes, the following information should be collected:

- Exact location and direction of the hole.

- Record of drilling progress.

- Core-losses, depth and extent, if possible also the causing factor.

- Depth at which water pressure is lost.

- Static water level, recovery and depth of hole whenever possible.

- Final depth of the hole.

A survey of the deviation of each hole from its intended course is usually required upon completion of the drilling. In order to obtain full hydrogeological and chemical information of a hole, all casing, grouting or supporting of the walls by concrete must be avoided, apart from the near-surface parts.

The investigations in these holes will generally follow the same pattern and procedures as in the shallow ones. Pumping will, however, only be made to obtain samples of the groundwater from selected sections isolated by straddle packers, whereas study of drawdown and outside response generally is not made. Flushing of the holes is made with nitrogen to protect the Eh of the water.

Detailed logging and laboratory examination of the drillcore, regarding mineralogy, chemistry, fracturing, fracture-filling, weathering and alteration is an important phase of the deep exploration.

The results of all these studies are evaluated and integrated with the results of earlier investigations to give a comprehensive characterization of the host rock, the hydrogeology and the geochemistry of the site.

5.5 Verification studies

The verification studies may comprise supplementary drilling in order to fill in details in that part of the site which has already been studied in the preceeding investigations, and also in order to probe other parts which were left out earlier. More important, however, they will comprise such often time-consuming investigations which serve to verify the assumptions and predictions of the hydrogeological and geochemical models developed for the site.

Such work may involve prolonged pumping in selected sections of the deep boreholes to obtain additional samples for chemical analysis and isotopic age determinations, as well as separate drilling to establish the natural stress situation in the host rock.

Other studies will serve to measure the hydraulic potential in different points at depth, and, if possible, to determine the volume of groundwater flow under natural gradients. Tracer tests to study the flow rate of groundwater and migration rate of different substances under artificial gradients between selected shallow drillholes are required at this stage to verify basic assumptions regarding the flow-porosity and the retardation effects of the host rock. Other critical issues for verifications may become evident only upon the conclusive analysis of the safety-determining factors for the site in question.

6. Acknowledgements

This presentation was originally prepared for the SwecoNuclear group of consultants. It is based on work for KBS of the Swedish Nuclear Fuel Supply Company, through the Geological Survey of Sweden. I would like to thank Mr. A. Göransson,

Mr. L. B. Nilsson and Dr. G. Kautsky, of these organizations, for granting permission for publication, and my colleagues at the Survey for helpful cooperation. The responsibility for the views expressed, errors and mistakes, rests entirely upon the author.

TECHNICAL BACKGROUND
Repository concept and basic design

Demands on site ↓ ↑ Demands on concept
Input data ↓ ↑ Natural conditions

SAFETY FACTORS
Safety analysis

Output data ↓ ↑ Input data
Demands on site ↓ ↑ Specifications met

SITE SPECIFIC DATA

Scope of investigations ↓ ↑ Synthesis of investigations

SITE INVESTIGATIONS
Results of all investigations

Figure 1. Diagram of information required to evaluate a given site

Table I. EXAMPLES OF INFORMATION THAT MAY BE DESIRED UNDER THE DIFFERENT HEADINGS OF FIGURE 1

TECHNICAL BACKGROUND
Nature and quantity of waste
Technical characteristics of repository
Nature of canisters, buffer and backfill
Thermal aspects
Geotechnical aspects

SAFETY FACTORS
Flow volume of groundwater contacting the waste
Concentration of nuclides in groundwater in contact with fissure minerals and host rocks
Lifetime of canisters
Transport times of nuclides versus depth of repository
Dilution of groundwater-transported nuclides before discharge at surface

SITE SPECIFIC DATA

Setting of the site

General description, landownership and use, accessibility and adjacent settlements
Regional geology, hydrology, morphology and seismicity

Conditions at the site

Local geology, hydrology, hydrogeology and geochemistry

Host rock conditions
Bedrock configuration
Main structures
Fracture frequency and orientation
Fracture history
Mineral composition, alteration, fracture minerals
Weathering
Geophysical maps and profiles
Drillhole evidence

Hydrogeology
Groundwater table and piezometry
Boundary conditions and locations
Groundwater temperature and density distribution
Hydraulic conductivities and diffusivities
Flow porosity
Flow model
Groundwater dating

Geochemistry
Analytical data on rocks, minerals, natural waters and gases
Eh, pH and ionic strength of waters
Nuclide-groundwater-mineral interaction
Mineral dating
Natural distribution of radioactive elements
Isotope data

Table II. SITE INVESTIGATIONS FOR A FEASIBILITY STUDY OF A HLW-REPOSITORY IN CRYSTALLINE ROCKS

A. PREPARATORY DESK WORK

1. Compilation of data

from available
literature and maps for
General description and
Regional setting of site
regarding
Landownership and use
Geography
Morphology
Hydrology (including data
on waterwells in bedrock)
Geology
Seismicity

2. Analysis of satellite imagery and airphotos

for
Fracture zones
Major structures
Rock units
Outcrops of bedrock
Topography
Drainage
Groundwater conditions

3. Analysis of airborne geophysics if available

Product:

Report with synthesis and maps
Suggestions for location of
geophysical traverses and
preliminary drillhole

B. PRELIMINARY TESTING OF SITE

1. Geological reconnaissance

Inspection of main outcrops
Field checking of satellite
airphoto and geophysical
interpretation

2. Geophysical traverses

by
Magnetometer
Slingram
Very Low Frequency VLF
to indicate hidden
Fracture zones
Dolerite dikes
Mineralizations

3. Test drillhole

500-1000 m deep, fully cored
to check rock conditions
at depth

Geological logging
Rock types
Rock quality
Fractures
Fracture fillings
Mineralization

Geophysical logging
Accoustic velocity
Caliper
Point resistance
Spontaneous polarization
Gamma radiation

Geochemical logging
Eh and pH, specific ions
Electrical conductivity of
water
Temperature of water

Product:

Go-no go decision for
continued work on site

Table II. SITE INVESTIGATIONS continued

C. SURFACE INVESTIGATIONS

1. Geological mapping

Rocks
Structures
Fractures

2. Hydrological mapping

Surface drainage
Springs
Waterwells

Determination of
Groundwater level
Temperature of water
Electrical conductivity, pH

Sampling of water
for chemical analysis

3. Geophysical mapping

by
Magnetometer
Resistivity
Induced polarization
Slingram and
Seismic refraction profiles

Interpretation aided by
Measurements on core samples
and geophysical logs of test
drillhole

Product:

Geological and geophysical
reports, maps and tentative
sections indicating locations
and inclinations of suspected
fracture zones
Suggested locations and
directions of shallow and
deep drillholes

D. SHALLOW EXPLORATION

1. Drilling of shallow holes

to about 100 m, directed at
suspected fracture zones

Sampling of well cuttings for
Geological logging
Chemical analysis
Determination of K_d values

Geophysical logging

2. Hydrogeological testing

Pumping tests
Drawdown and recovery
Response in adjacent wells
Sampling of water for
chemical analysis

Injection tests

Tracer tests

3. Geological, geochemical and
hydrogeological evaluation

Product:

Tentative 3-D model of hydrogeological conditions at the
site, with location, direction
and near-surface characteristics
of main waterbearing zones
Revised locations and directions
for deeper drillholes

Table II. SITE INVESTIGATIONS continued

E. DEEP EXPLORATION

1. Drilling of 3-10 deep holes

500-1000 m deep, fully cored

2. Geological, geophysical and geochemical logging

3. Hydrogeological testing

Injection tests
Sampling of water and gases for analysis and age determination
Piezometry, tidal effects
(Dilution and tracer tests)

4. Auxiliary study of cores

Characterization of rocks and fracture zones
Physical and geotechnical properties
Mineralogy
Geochemistry

5. Auxiliary study of drillholes

TV-inspection
Electrical investigations (Mise-à-la-masse)
Electromagnetic studies
Seismic studies

6. Evaluation

Processing of data
Analysis and interpretation
Synthesis
Geological model of site
Hydrogeological model
Numerical analysis of groundwater flow
Geological prediction

Product:

Functional description of site
Input to safety analysis and grading of site
Plan for verification studies

CURRENT SITE-INVESTIGATIONS FOR DEEP GEOLOGICAL DISPOSAL IN SWEDEN

S. Scherman and N. Rydell
National Council for Radioactive Waste
Stockholm, Sweden

ABSTRACT

The Geological Survey of Sweden has on comission from the National Council for Radioactive Waste carried out bedrock investigations in Sweden during the last 3 years. The investigations are of a generic character and made in commonly occuring crystalline rocks. Their purpose is to obtain characteristic data on petrology, groundwater, chemistry, tectonics and permeabilities down to depths of 500 to 800 m.

Results are presented from mainly two areas each of approximately 4 km^2 each.

The work so far completed involves detailed geologic and tectonic mapping as well as interpretation of data and observations from airphotos and geophysical measurements.

INTRODUCTION

The research and development work in Sweden on high-level waste disposal has as one of its objectives to identify and characterize geologically suitable host rocks for a HLW-repository. The scheduling of this research work is set by the time schedule for the repository which should be in operation about the year 2020. The licensing of a site and the construction of the repository and the auxiliary facilities may take as much as 20 years. This still leaves 20 years for geologic investigations leading up to proposals for one or more sites with bedrock well suited for a HLW-repository. The work is divided in two phases. The first phase, during the 80-ies, is mainly devoted to generic bedrock investigations. 8 to 10 sites exemplifying commonly occuring crystalline rock types will be studied. Experience from this work as regards rock characteristics as well as investigation techniques will successively be introduced in the direct search for suitable disposal sites which will be the main objective of the second phase of the work.

These investigations are made by the Geological Survey of Sweden on contract from the National Council for Radioactive Waste, a government agency which supervises and initiates R and D work on radioactive waste management.

BACKGROUND AND EARLIER EXPERIENCES.

The bedrock in Sweden is composed of hard crystalline rocks with ages varying from 2000 - 1000 million years. Younger, sedimentary rocks (shale, sandstones etc.) are only to be found in smaller limited areas.

Extensive experiences from underground construction in hard rock have contributed knowledge on different rock types about hydrology, rock mechanics etc.

The following properties can often be regarded as less suitable for a repository host rock

o Intense fracturing

o Strongly schistozoic structure

o Lineated structure with open fractures

o Clay alterations within the rock-mass

o Secondary red-colouring

o Sharp changes in rock composition and rock types.

Considerations of this nature guided the first choice of investigation areas during the KBS-project.

The following bedrock types were studied

o granodiorites in the middle east of Sweden (Finnsjön)

o silica rich young granites in southeastern Sweden (Kråkemåla)

o sedimentary gneisses in southern Sweden (Sternö)

The results from these investigated areas show quite clearly that the main surface structures can be followed down to depth except for rock formations with horizontal fracturing, e.g. the granite in Kråkemåla.

These studies also indicate that the structure of the bedrock is of great importance.

CURRENT VIEW ON GEOLOGIC CRITERIA.

Through studies of literature, inverviews with underground constructors, hydrologists and structural geologists we came to the following conclusions:

Homogeneous granites often show a pronounced, well defined and pervading fracture pattern, usually in three perpendicular directions, which results in high permeability along fractures. The homogeneous compositon and well defined fractures makes it on the other hand easier to predict real conditions down to depth than in more structured rock types.

Gneissic or structured rock types such as migmatites and veined gneisses show quite different characteristics compared to the above mentioned granites.This can be explained by the plastic folding of these rock types which often causes the planes of schistosity to change direction and therefore counteracts pre-vading fracturing (irregual fracture pattern).

A great number of data on water capacities in rock drilled wells support these observations.

Our "site selection criteria" are based on such experiences.

SELECTION CRITERIA FOR EXPLORATORY SITES.

The following criteria are used when areas are selected for more detaliled geologic and hydrologic investigations.

- o The rock types shall have gneissic or at least structured character or otherwise high competence against deformation
- o Low frequency of regional chrushed zones
- o Low fracture frequency within exposed bedrock
- o The composition and structure of the rock mass shall be as uniforme as possible
- o Data from rock-drilled wells shall indicate low groundwater flow.

METHODOLOGY

Studies of geological maps, data from airborne geophysical measurements, sattelite photo and airphoto interpretation in scale 1:60 000 as well as data from water capacities in rock drilled wells are used in the first selection of greater bedrock areas of interest.

These informations are plotted on topographical maps in scale 1:50 000.

Exploratory field investigations including generalized bedrock and fracture mapping are made for each area in order to verify the results from the desk studies.

One or two areas, each of approximately 4 km^2, are then selected for further investigations including

- o airphoto interpretation in scale 1:30 000
- o detalied bedrock mapping in scale 1:2000 or 1:5000
- o fracture mapping including orientation, infillings, widths, lengths, statistics etc. in scale 1:2000 and 1:5000
- o surface geophysics including resistivity-, magnetic-, IP-, VLF- and slingram measurements.

A drilling program is then specified. Drillholes are positioned and directed towards identified anomalies at desired depths and through undisturbed portions of the bedrock. 5 to 6 drillholes with a total length of 3000 to 4000 m will normally be sufficient to test the results of surface geophysics and to give a good overall characterisation of the rock body. Shallow percussion drilled holes may be used as pilot holes in order to verify predicted dips and strikes of anomalies before the deep core holes are positioned. The cores are mapped in detail and the rock mass explored with geophysical bore hole equipment. Ground water is sampled at different depths for chemical analyses and age determinations. Permeabilities are measured.

SELECTED SITES

Four areas of 4 km^2 each have so far been studied

- o sedimentary gneisses in Blekinge, southern Sweden (geologic and tectonic mapping)
- o veined sedimentary and granitic gneisses in Bohuslän on the Swedish west-coast (geologic and tectonic mapping and surface geophysics)
- o migmatized gneissic-granitic rocks in Hälsingland in central Sweden (geologic and tectonic mapping, airborne and ground surface geophysics)
- o gabbromassif in Lappland in northern Sweden, exploratory field mapping, airborne magnetic and gravimetric measurements)

Deep investigations, except for one coredrilled hole in Hälsingland, have been delayed by interference from local opposition groups.

This report will therefore concentrate on the results from surface mapping and geophysical measurements where these have been completed i.e. in Hälsingland and Kynnefjäll (Bohuslän).

FIELD INVESTIGATIONS

Veined Gneisses, western Sweden (Kynnefjäll).

General

The investigated area forms part of a topographically distinguished high plateau. The plateau lies appr. 160 m above sea-level with the surronding valleys 50 meters above sea-level. The plateau is 200 km^2 in size and the site area 5 km^2. The bedrock is well exposed.

Bedrock geology

Two dominating rock types are defined

- o veined gneisses of sedimentary origin(eastern part of the area) with Quarz, Feldspar, Biotite and Muscovite
- o gneissgranite (western part of the area) with Quarz Feldspar and Biotite.

Veins of amfibolite occur partly.

The geochemical composition of these rock types can be read from table I.

TABLE I

Chemical analyses from Kynnefjäll (% of weight).

Rock type	Amfibolite	Gneiss-granite	Gneiss	Gneiss-granite	Gneiss
SiO_2	57.0	71.0	65.2	67.2	72.2
TiO_2	0.94	0.56	0.62	0.79	0.51
Al_2O_3	14.0	14.6	17.2	15.6	14.1
Fe_2O_3	1.9	0.7	1.7	0.7	0.5
FeO	9.9	2.9	3.8	3.9	3.1
MnO	0.23	0.06	0.09	0.09	0.07
CaO	8.6	0.6	1.0	3.5	1.4
MgO	5.7	1.46	1.92	1.64	1.34
Na_2O	1.5	1.7	2.1	3.6	2.6
K_2O	0.8	6.8	5.2	2.6	3.8
BaO	0.03	0.14	0.11	0.06	0.08
$H_2O > 105^o$	1.1	1.2	1.8	0.8	1.0
$H_2O < 105^o$	0.1	0.1	0.2	0.1	0.1
P_2O_5	0.03	0.02	0.04	0.22	0.01
CO_2	0.05	0.06	0.10	$<$0.01	0.02
F	0.04	0.05	0.07	0.07	0104
S	$<$ 0.02	$<$0.02	$<$0.02	$<$0.02	0.02
Total	101.92	101.95	101.15	100.87	100.89

The gneisses form the oldest rock which has later been intruded by the granites. The contact zone, 20-100 meters broad, striking approximately N-S, is diffuse and shows no signs of tectonic activity. The dip of the contact has not been possible to verify but it probably dips 20-60° to the east.

Both rock types have a pronounced schistosity striking approximately N-S with shallow dips to the east.

Fracture tectonics

From airphoto investigation and from the field survey it is obvious that the area is penetrated by several fracture zones. Most of them are small and weakly indicated in the terrain. The dominating strikes are N-S and NE-SW. These directions correlate well with the regional lineament pattern.

Observations in the field indicate steep dips which in some cases have been verified by geophysical measurements.

Small scale fracturing is mainly steep and striking E-W.

Geophysical Measurements

A combination of several geophysical methods were used in order to obtain information on different physical parameters in the bedrock.

The following methods were used.

o magnetic

o slingram

o resistivity and induced polarization

o VLF (exploratory)

The results are presented as computer generated maps in scale 1:5000.

A comparison between the geologic and tectonic field mapping and the results from the geophysical measurements shows good correlation.

Migmatite, central Sweden.

General

The area is situated at approximately 200 meters altitude above sea-level. The hill crest which defines the site is surrounded by valleys in NW direction. The surface is motly covered with moraines and the bedrock is sparsely exposed.

Bedrock geology

Migmatized sedimentary rocks dominate. The presence of garnet, sillimanite and cordianite indicates an argillitic origin. The structure is pronounced, striking to the NW. Due to the low frequency of outcrops, results from airborne magnetic measurements had to be used to locate the migmatite, which has a lower degree of magnetization then the surrounding granites.

Fracture tectonics

The regional fracture pattern, as seen from airphotos, indicates dominating strikes in NW and WNW directions while the electrical airborne measurements show low resistivity lines predominnatly in NE direction.

Field mapping and surface geophysical measurements are being evaluated at present and will be presented at the workshop together with a summary of the results from on corehole down to 800 m depth.

Gabbromassif in Lappland, northern Sweden.

The program of the National Council on bedrock investigations suggests that one exploratory site shall be situated in a gabbromassif.

The following properties of gabbro justify this choice

- o high competence against deformation
- o due to the geochemistry it easily forms fracture minerals, which will decrease the water conductivity
- o low water capacity in rock drilled wells compared to other hard rocks (e.g. granites)
- o due to high density and high magnetic properties it is easily mapped by geophysical methods
- o gabbromassifs of extensive volume occur frequently in northern Sweden
- o a considerable amount of background data fram Swedish gabbros is available.

A pre-investigation study of available geophysical material resulted in four suggested exploratory sites.

Two criteria were used for the final selection

- o the gabbro shall reach a depth of at least 1000 meters
- o the fracturing or the deformation of the massif shall be as low as possible.

From airphotos, geological maps and finally field exploratory work, one area (Taavinunnanen) was chosen for extensive studies and core drillings.

Area description and short rewiev of geophysical data.

The Taavinunnanen gabbro is situated above the treelimit NE of the town of Kiruna. The gabbro has an elliptical form with the major axis oriented N-S and covers an area of approximately 50 km^2. The rock exposure is good.

The gabbro intrusion is well distinguished on the magnetic map and also shows a magnetic layering dipping towards the center of the gabbro.

On the gravity map the gabbro shows a positive anomaly due to its higher density compared to the surrounding rocks (gneisses and granites). The depth of the gabbro has been calculated from this anomaly to be approximately 2000 meters.

Future work within this area will start during the summer 1981 with one exploratory core drilling in the center of the massif down to approximately 700 meters and detailed geologic and tectonic mapping.

DISCUSSION

M.F. THURY, Switzerland

Vous avez trouvé dans les quatre régions étudiées différents degrés de fracturation. Avez-vous déjà des informations ou des hypothèses sur l'influence du uplift isostatique postglaciaire sur les fracturations ou les réouvertures de fracture ?

S. SCHERMAN, Sweden

No evidence have so far been found which shows reactivation or opening of fractures due to rebound of the crust in our selected sites. A programme carried out by the Geological Survey of Sweden, will start during the summer-81 aiming to a better understanding of how seismic and isostatic activity influence a bedrock formation.

E. GOSK, Denmark

Do you have any data available from the pumping tests you have performed, particularly the relation between fissure distribution of permeability ?

K. AHLBOM, Sweden

Referring to water capacity in rock drilled wells, data are available to a certain extent.

Work on the relation between fissure distribution on a regional scale and water capacity from rock drilled wells has not been done.

EARTH-SCIENCE DATA NEEDS FOR SITE CHARACTERIZATION

S. S. Smith
Office of NWTS Integration, Battelle Memorial Institute
Columbus, Ohio (United States)

O. E. Swanson
Office of Nuclear Waste Isolation, Battelle Memorial Institute
Columbus, Ohio (United States)

ABSTRACT

The site screening activities of the United States NWTS program are narrowing down to several sites throughout the U.S. As a result, considerable effort is currently being channeled into the determination of the data that must be obtained to fully characterize a specific site in order to construct, license and operate a repository at the site. The Office of NWTS Integration (ONI) has recently orchestrated the development of a generic document that discusses the earth-science data to be aquired during site characterization activities and the preferred method(s) of acquiring this data. This paper discusses the general steps involved in the development of the subject document and summarizes the data needs presented in the document pertaining to the earth-science area of geology and geohydrology.

The general site screening and selection process utilized by the United States, Department of Energy (DOE) for the siting of nuclear waste repositories in geologic formations was outlined in a previous workshop paper by Newcomb, et al. [1]. The national effort is designated the National Waste Terminal Storage (NWTS) program. In recent years the NWTS program has been conducting general site screening activities throughout the contiguous U.S. The investigations performed as part of the screening process have included general reviews of existing geologic information, reconnaissance field work and geologic mapping, geophysical surveys and exploratory drilling. The data obtained from these investigations was, by planning, only utilized to screen sites, not characterize a site. However, now that the NWTS program is narrowing down to several sites throughout the U.S., considerable effort is being channeled into the determination of the data that must be obtained to fully characterize a site in order to construct, license and operate a repository. Three site characterization plans are currently being developed within the NWTS program: salt domes in the gulf coast interior region; the DOE Hanford Reservation in the State of Washington; the DOE Nevada Test Site in the State of Nevada.

The Office of NWTS Integration (ONI), Site Program Office has recently orchestrated the development of a generic document that discusses the earth-science data to be acquired during site characterization activities and the preferred methods(s) of aquisition of this data. The title of the document is "Earth-Science Data Acquisition Guidelines" (ESDAG). This document is intended, among other things, to serve as a guideline during the preparation of site specific characterization plans. In the context of this document, earth-science is represented by programmatic areas of geology, geohydrology, geochemistry, rock mechanics and soil mechanics. A preliminary draft of ESDAG is currently being reviewed by the NWTS community. A document addressing the environmental data to be obtained during site characterization is also being prepared.

This paper discusses the general steps involved in the development of ESDAG. Space limitations permit the presentation of only a fraction of ESDAG's contents. For reasons discussed subsequently the data needs and aquisition methods developed for the earth-science areas of geology and geohydrology are summarized herein.

Programmatic consensus of the earth-science data that must be aquired to properly characterize a repository site has been an illusive goal within the NWTS program. The size and diversity of the NWTS program has been a major contributing factor to the inability to reach a consensus. The DOE has several prime contractors managing and performing site screening activities in many geologic media (i.e., domal salt, bedded salt, granite, basalt, tuff) at scattered land units throughout the U.S. (e.g., States of Utah, Texas, Mississippi, Washington, etc.). There are many specialty areas within the program that will require earth-science data. As a consequence, there are many varied uses for the same data and conversely, there are unique earth-science data needs required by some of the specialty areas.

In order to assure technical and programmatic interface through the massive NWTS program, DOE created ONI in mid-summer 1980. One of the first tasks assigned to ONI was the development of the earth-science data that should be acquired to adequately characterize a repository site; that heretofore illusive goal. As a first step ONI determined the technical, specialty areas within the NWTS program that will require earth-science data to achieve their programmatic goals. The specialty areas included geologists, geohydrologists, seismologists, repository designers, waste package designers, penetration sealing designers, geochemists and performance assessment modelers. A task group consisting of one to two representatives from each specialty areas was formed. The representatives were selected by way of their technical qualifications and their familiarity with the NWTS program.

The justification for earth-science data needs should be properly structured and tied into existing program baseline documents. A shotgun approach of simply listing the data needs that "come to mind" is unacceptable in such a complex, multidisciplinary program. The NWTS-33 series of documents was utilized to properly baseline the efforts of the Task Group. These documents present performance criteria that designate requirements for the performance of the disposal system and its components (i.e., site, respository, waste package).

"Site Performance Criteria" NWTS-33(2) [2] was recently published by DOE. This document was the primary baseline document employed by the Task Group. NWTS-33(2) presents criteria and subcriteria that DOE is using to screen sites and to evaluate the suitabilty of sites. The first five criteria cover the earth-sciences and the last five criteria cover human intrusion, surface characteristics, demography, environmental protection, and socioeconomic impacts. The five criteria and associated subcriteria pertaining to the earth-sciences are reproduced in Figure 1. During site screening activities the ten criteria are selectively applied in the search for potentially suitable sites. The criteria are then (during the site suitabilty evaluation) used to determine the suitability of specific sites identified during site screening.

The following approach was employed to generate the preliminary draft of ESDAG. Each specialty area representative(s) determined the analyses, evaluation, assessment, etc., that must be performed in his or her area to assure the site fulfills the applicable five earth-science related criteria and subcriteria contained in NWTS-33(2). The earth-science data necessary to perform the analyses and the preferred field and/or laboratory method(s) used to acquire the data were then evaluated. It was assumed for the purpose of ESDAG that all data would be obtained from surface-based exploration.

Geology and geohydrology earth-science data were basic requirements of all of the specialty areas. For this reason and because of space limitations only these data needs developed by the Task Group are reproduced herein. One must refer to the original document to extract the Task Group's summary of data needs in the areas of geochemistry, rock mechanics and soil mechanics. Table 1 summarizes the data item, use of the data and preferred methods(s) of acquisition.

Table 1 lists more than one method of acquisition for many of the data items. It is not intended that all methods necessarily be used. The final selection of a method or methods is dependent upon the particular site and subsurface media being characterized.

The geology and geohydrology data listed in Table 1 is not intended to serve as a shopping list of data that must be amassed at every site. Actual data requirements should be evaluated on a site-specific basis and summarized in a Site Characterization Plan. As an illustration, possible volcanism is a key issue in the siting activities at the DOE Nevada Test Site. The data items pertaining to volcanism listed in Table 1 would be germane to site characterization activities at the Nevada Test Site. In contrast, salt dome site characterization activities in the interior Gulf States would expend little effort, if any, collecting data pertaining to volcanism.

As discussed above, a preliminary draft of ESDAG is currently being reviewed by the NWTS community. The NWTS program welcomes a critical review by the NEA/OECD of the geology and geohydrology, site characterization data needs and acquisition methods listed in Table 1. It is felt that site characterization data needs concurred upon in an international forum would enhance the credibility of future site characterization plans and activities.

The earth-science data that will be collected in subsequent site characterization activities will support regulatory reviews. In the case of the U.S., the data will be reviewed by the Nuclear Regulatory Commission in their assessment of the suitability of a site for the granting of a license. In anticipation of these reviews the NWTS program is querying itself about the level of quality the data should have. The questions that come to mind are:

- What management procedure could be utilized to assess if existing, standardized laboratory and field test procedures provide the necessary precision and quality of data for licensing?

- Given the redundancy inherent in the field and laboratory testing program, is it necessary to even address the question about the precision of a particular test result, assuming the test procedure followed was accepted as adequate by the appropriate technical specialists?

It is hoped that we may elicit your input in response to these questions in the course of this workshop.

BIBLIOGRAPHY

(1) Newcomb, W. E., Smith, S. S. and Swanson, O. E., "Site Selection in the National Waste Terminal Storage Program: An Overview of Technical and Political Aspects of the Problem," Proc. Workshop on Siting of Radioactive Waste Repositories in Geological Formations, NEA/OECD, Paris, May, 1981.

(2) "NWTS Program Criteria For Mined Geologic Disposal of Nuclear Waste: Site Performance Criteria", National Waste Storage Program, U.S. Department of Energy NWTS Program Office, Columbus, Ohio, DOE/NWTS-33(2), February, 1981.

Figure 1

NWTS SITE PERFORMANCE CRITERIA PERTAINING TO THE EARTH-SCIENCES

1. Site Geometry

The site shall be located in a geologic environment that physically separates the radioactive wastes from the biosphere and that has geometry adequate for repository placement.

(1) The minimum depth of the repository waste emplacement area shall be such that credible human activities and natural processes acting at the surface will not unacceptably affect system performance.

(2) The thickness and lateral extent of the geologic system surrounding the waste emplacement area shall be sufficient to accommodate the repository and a buffer zone and to ensure that impacts induced by construction of the repository and by waste emplacement will not unacceptably affect system performance.

2. Geohydrology

The geohydrologic regime in which the site is located shall have characteristics compatible with waste containment, isolation, and retrieval.

(1) The site shall be located so that the present and probably future geohydrological regime will minimize contact between ground water and wastes and will prevent radionuclide migration or transport from the repository to the accessible environment in unacceptable amounts.

(2) The site shall be located so that the hydrological regime can be sufficiently characterized to permit modeling to show that present and probable future conditions have no unacceptable impact on repository performance.

(3) The site shall be located so that the geohydrological regime allows construction of repository shafts and maintenance of shaft liners and seals.

(4) The site shall be located so that subsurface rock dissolution that may be occurring, or is likely to occur, can be shown to have no unacceptable impact on system performance.

3. Geochemistry

The site shall have geochemical characteristics compatible with waste containment, isolation, and retrieval.

(1) The site shall be located so that the chemical interactions between radionuclides, rock, ground water, or engineered components will not unacceptably affect system performance.

4. Geologic Characteristics

The site shall have geologic characteristics compatible with waste containment, isolation, and retrieval.

(1) The site shall be located so that the subsurface setting can be sufficiently characterized to permit identification and evaluation of conditions that are potentially adverse or favorable to waste containment, isolation, and retrieval.

(2) The site shall provide a geologic system which can be shown to accommodate anticipated geomechanical, chemical, thermal, and radiological stresses caused by waste/rock interactions.

Figure 1
(Cont.)

---

(3) The site shall be located so that development, operation, and closure of underground areas can be accomplished without undue hazard to repository personnel.

5. Tectonic Environment

The site shall be located such that credible tectonic phenomena will not degrade system performance below acceptable limits.

(1) The site shall be located so that its tectonic environment can be evaluated with a high degree of confidence to identify tectonic elements and their impact on system performance.

(2) The site shall be located so that Quaternary faults can be identifed and shown to have no unacceptable impact on system performance.

(3) The site shall be located so that the centers of Quaternary igneous activity can be identified and shown to have no unacceptable impact on system performance.

(4) The site shall be located so that long-term, continuing uplift or subsidence rates can be shown to have no unacceptable impact on system performance.

(5) The site shall be located so that ground motion associated with the maximum credible earthquake will not have unacceptable impact on system performance.

Table 1

GEOLOGY AND GEOHYDROLOGY DATA NEEDS

GEOLOGY/REGION		
Data Item	Use of Information	Preferred Method of Data Acquisition
Historic Sea Level Data	Evaluation of possible future erosion and or denudation	Literature review of published/unpublished geologic data
History of Glaciation	See Above	Literature review of deep sea core sample data
Local Pleistocene and Holocene Stratagriphic Record	Evaluation of possible future erosion and loading by glaciation	Literature review of published/unpublished geologic data; surface geologic mapping; logging of trenches; coring and logging of surficial materials; and age dating
Geothermal Gradient	Assessment of present or possible future geothermal or volcanic conditions, established ambient temperature conditions	Temperature logging in deep exploratory well
Thermal Conductivity of Rock	See Above, performance of thermomechanical analyses	Laboratory testing of core sample
Thickness of Overlying Strata or Surficial Deposits	Evaluation of depth to basement rock	Drilling and sampling of deep exploratory wells; seismic reflection or refraction surveys; gravity survey
Map of Faulting, and History of Offset	Characterization and location of faulting and rifting, evaluation of seismic design hazard evaluations and development of design earthquakes, evaluation of effect of faulting on geohydrologic regime and setting	Geologic field mapping; review of existing well records; seismic reflection surveys; logging and core sampling of deep exploratory well; logging and sampling exploratory trench; age dating of datable sampled materials
Nature of Fault Cross-Cutting Relationship	Determination of sense of faults and history of offset	Review of existing well records; seismic reflection surveys; logging and core sampling of deep exploratory well
Subsurface Location of Fault Plane	Location of faulting	Review of existing well records; seismic reflection surveys; logging and core sampling of deep exploratory well

Table 1
(Cont.)

GEOLOGY/REGION (Cont.)

Data Item	Use of Information	Preferred Method of Data Acquisition
Location and Magnitude/ Intensity of Historical Earthquake Events (Preinstrumental/instrumental)	Evaluate the presence and level of earthquake activity, assist with evaluating existence of active earthquake fault or feature, evaluation of general seismicity, development of design ground motion parameters	Review of earthquake literature; macro/ microearthquake monitoring network at/near site
Accelerograph of Recorded Earthquakes	See Above	See Above
Principal Stress, Orientation and Magnitude	Knowledge of stress regime can be used to evaluate regional tectonic compression or extensor characteristics, also used in design of underground openings and bracing	Hydrofracture tests performed in deep exploratory well holes; microearthquake monitoring network; In situ stress measurement methods performed in deep exploratory well holes
Outcrop Map of Existing Volcanoes	Evaluation of possible present and future volcanic effect on site	Literature review of published/unpublished data
Map of Flows and Debris Fallout	See Above	Literature review of published/unpublished data
Chemical and Petrographic Characteristics of Previous Flows and Debris Fallout	See Above	Laboratory chemical tests of rock samples; (i.e., thin section analysis, X-ray difraction, mass spectroscopy, probe data)
Elevation Changes	Evaluation of possible and future uplift or subsidence and resulting effect on erosion/denudation and geohydrologic regime and setting.	Geodetic survey and resurveys
Bouger Anomaly Map	See Above	Bouger Gravity Surveys
Evidence of Previous Ice or Water Cover	See Above	Literature review of published/unpublished data

Table 1
(Cont.)

SITE GEOLOGY/OVERBURDEN AND HOST ROCK		
Data Item	Use of Data Information	Preferred Method of Data Acquisition
Sequence and Thickness of Subsurface Units (Including Aquifers and Aquitards)	Characterization of subsurface geologic and geohydrologic setting; data are used by repository designer, performance assessment modeler, sealing penetration designer	Review of existing wellhole records; logging and sampling of deep exploratory wells (calibration holes); downhole geophysical logging methods; surface seismic reflection profiling
Lateral Extent of Subsurface Units (including Aquifers and Aquitards)	See Above	See Above
Location and Dimensions of Facies Changes, Voids, Lenses, etc.	See Above	See Above
Orientation, Spacing, Persistence, Aperture, Openness and Surface Roughness of Fractures, Joints, and Foliations	See Above	Logging and sampling of deep exploratory wells; surface geologic mapping, sensing imagery; oriented rock core samples from exploratory wells; downhole camera; impression packer
Location, Attitude and Filling Mineralogy of Faults/Shears	See Above	See Geology/Region, map of faulting and subsurface location of fault plane; and petrographic analyses of samples of discontinuity (thin section, X-ray diffraction, mass spectroscopy; probe data)
Location, Amplitude and Wavelength of Folds	See Above	Review of existing wellhole records; logging and sampling of deep exploratory wells (calibration holes); downhole geophysical logging methods; surface seismic reflection profiling
Size and Shape of Diapirism	See Above	Gravity Survey and see above
Composition and Texture of Rock	See Above	Petrographic analyses of core samples or cuttings (thin section analyses, X-ray diffraction, mass spectroscopy, probe data)

Table 1
(Cont.)

SITE GEOLOGY/OVERBURDEN AND HOST ROCK (Cont.)		
Data Item	**Use of Data Information**	**Preferred Method of Data Acquisition**
Depth to Bedrock	Characterization of subsurface geologic and geohydrologic setting; data are used by repository designer, performance assessment modeler, sealing penetration designer	Surface geologic mapping; surface seismic refraction survey
Location and Extent of Nonanomalous Areas	See Above	Petrographic analysis of rock samples; surface seismic refraction survey; ground and aerial, magnetics and gravity survey, dipole-dipole resistivity, audiomagneto tellurics
Disturbed Zone and Interface Zone Permeability of Drill Holes and Shafts	Influences seal zone permeability; needed to determine requirement for and design of disturbed zone cutoff sealing device	Guarded straddle packer hydrologic drill stem test; laboratory tests in association with well bore damage simulation
Size, Shape and Deviation of Drill Hole and Shaft Penetration	Determine seal shape and volume for penetration seal design	Downhole: caliper log, T.V. camera, and seisviewer
Penetration (drill hole and shaft) Deformability	Determine long-term seal-host rock mechanical interaction for penetration seal design	Monitoring of borehole closure; in situ deformation jack; lab test in conjunction with well bore damage simulation

Table 1
(Cont.)

GEOHYDROLOGY		
Data Item	**Use of Data Information**	**Preferred Method of Data Acquisition**
Potentiometric Head of Aquifer[1] and confining Units	Evaluation of piezometric surface, velocity and direction of flow and location of recharge/discharge areas	Hydrologic drill stem (single and straddle packers); installation of open tube, pneumatic, hydraulic, or electric piezometer in deep well
Formation Fluid[1] Pressure		
Porosity or Void[1,2] Ratio of Aquifers and Confining Units	Numerical groundwater modeling	Select downhole geophysical logging methods (see Appendix A); laboratory testing of rock core samples or undisturbed soil samples
Permeability of[1,2] Aquifers and Confining Units (Vertical; Horizontal)	Evaluation of transmissivity and velocity of flow; numerical groundwater modeling	Hydrogeologic drill stem testing in exploratory boring or deep well; laboratory testing of rock core sample or undisturbed samples of soil; field tracer tests; falling head, recharge or constant head tests in shallow unconfined aquifers
Groundwater Age of[1,2] Aquifers	Evaluation of rate of movement of groundwater from recharge area	Laboratory age-dating of groundwater sampled from exploratory borings or deep wells
Groundwater[1,2] Temperature	Needed as input to groundwater models (thermal potential)	Temperature logging in exploratory borings or deep wells
Groundwater[1,2] Viscosity and Density	Needed as input to groundwater models	Laboratory testing of groundwater sampled from exploratory borings or deep wells
Groundwater[1,2] and Aquifer Compressibility	Data required for groundwater models	Laboratory testing of groundwater samples and undisturbed soil samples or rock core samples
Water Table[2] Elevation	Data required for modeling	Logging and sampling shallow test well; installation of piezometers; surface electrical resistivity survey; water-table hydrograph measured in exploration hole.

Table 1
(Cont.)

GEOHYDROLOGY (Cont.)		
Data Item	Use of Data Information	Preferred Method of Data Acquisition
Transmissivity, Storage coefficient, specific capacity	Design of water control systems; numerical groundwater modeling	Pump tests in wells
Groundwater Recharge points (each bed)	Design of water control systems; numerical groundwater modeling	Piezometers; aerial photos; well logs
Groundwater Discharge Points (each bed)	Radionuclide migration predictions, numerical groundwater modeling	See above

[1]Confined; [2]Unconfined

POSSIBILITY OF INFERRING SOME GENERAL FEATURES AND MINERALOGICAL COMPOSITION OF DEEP CLAY BODIES BY MEANS OF SUPERFICIAL OBSERVATIONS^

B. ANSELMI, A. BRONDI, O. FERRETTI, V. GERINI
CNEN (Comitato Nazionale Energia Nucleare)
ITALY

ABSTRACT

The CNEN (Italian Nuclear Energy Commission) is highly engaged in the study of many physical features of the territory for sitological purposes. In this frame the deeply buried clay deposits represent an area of study of great interest. Direct informations on deep deposits are often lacking. The CNEN has therefore faced the problem of the possibility of superficial observations to be used in predicting some characters of underground clay bodies. Systematic investigations carried on pliocenic clays occurring in Italy have shown: 1)Pliocenic clay deposits show a clear regional distribution according well defined mineralogical provinces; 2) Mineralogy of clay deposits coarsely depends on lithological composition of ancient feeder basins. The obtained results may allow extrapolations to deep deposits.

RESUME

Le CNEN (Comité National pour l'Energie Nucléaire) a entrepris une étude du territoire italien en ce qui concerne les sites. Les dépôts argileux profonds étudiés dans la présente communication semblent être un domaine intéressant et sur lequel il y a généralement peu d'informations directes. Le CNEN a étudié la possibilité d'utiliser des observations faites en surface pour préjuger des caractéristiques des formations argileuses en profondeur. Les investigations systématiques conduites sur les argiles pliocéniques ont montré que 1) les dépôts pliocéniques ont une distribution régionale très nette qui correspond à des provinces minéralogiques bien déterminées ; 2) la minéralogie des dépôts argileux dépend grossièrement de la composition lithologique des anciens bassins d'alimentation. Ces résultats peuvent s'étendre aux dépôts profonds.

^ Work supported by the Commission of European Community.

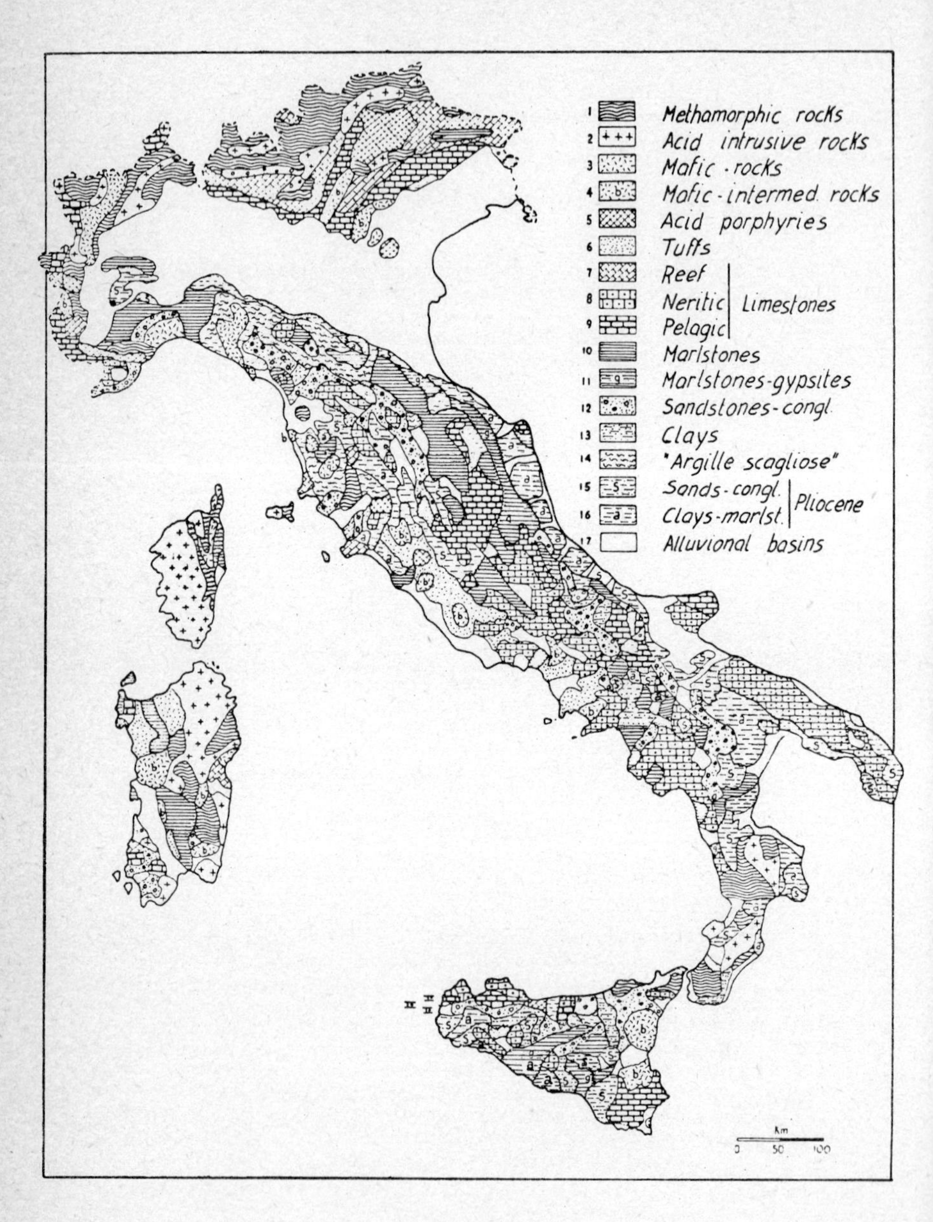

Fig. 1 Simplified lithological map of Italy.

The CNEN (Italian Nuclear Energy Commission) is presently developing researches aimed to the qualification of some geological formations for a long time isolation of nuclear waste. These researches
are conducted in a more general frame of investigations coordinated
and in part financially supported by the European Community. Granit,
clay and salt are at the moment the three geological formations studied by the Member States of the Community. Belgium and Italy are engaged in the study of the clay formation.

As for the other mentioned formations a catalogue has been com
piled by the Community identifying clay deposit showing suitable con
ditions of extention, thickness and deepness for the disposal of nuclear waste. Other important parameters must be therefore acquired
in order to definitively ascertain the suitability of the retained de
posits. First of all homogeneity thickness and vertical and lateral
continuity of the deposits must be considered. Mineralogical composition, exchange capacity, geotechnical characteristics and other aspects are however of not negligible importance. Taking into account
that these informations are lacking for all or almost all the deposits considered in the catalogue the problem is to verify if some indications may be obtained by means of indirect observations before
using expensive operations of superficial geophysical surveys and deep
drillings. In order to explore the real possibility and usefullness
of this kind of indications systematic researches has been conducted
on the outcropping marginal parts of the most important clay deep deposits of pliocenic age existing in Italy. The main achieved results
are here reported together to some theoretic considerations regarding
some general structural features of clay deposits according to physico-
geographic evolution.

The geology of Italy is very complicated (fig. 1). Starting
from Miocene an intense sedimentation took place in syn- and post-oro
genic phase. The sedimentary basins of the italian peninsula may be
grouped in two fundamental types: a) basins open seaward, as in the
eastern side; b) marginal and internal basins, mainly occurring in
the western side.

As it is possible to note in the geologic map, the various sedimentary basins may have been supplied with sediments originated by
very different parent rocks. This fact may have caused: 1) very varia
ble volumetric, areal and stratigraphic relationships between sand
sediments and clay sediments accumulated in the different basins;
2) important variations in the mineralogic composition of the clay
deposits. If this picture would correspond to a real situation it
could be considered an usefull tool both for a coarse predicting of
the main characters of deep clay deposits and in orientating expensi
ve operations such as deep drilling. In order to give a first approximation answer to this problem the pliocenid clay deposits occurring
in Italy have been sistematically sampled. Further investigations are
planned for the examination of clay deposits of different age.

About 70 clay samples have been collected. Clay deposits belonging to seaward open basin have been sampled in their peripheric outcropping parts. The obtained informations must therefore be considered as corresponding to an edge character. The samples may on the con
trary correspond also to the bulk of the deposits in the case of marginal basins (fig. 2).

The collected samples have been submitted to granulometric analysis. The sand fraction ($>$ 0.06 mm) has been sieved at 1/2 phi in-

tervals. The clay fraction has been examined by means of hydrophotometer. Figure 3 reports the granulometric histograms of the more representative samples. As it is possible to note the samples of the various zones don't show significative granulometric differences and the silty fraction is important in all the considered cases. This fact may derive by the edge conditions of the sampled deposits.

Diffractometric analysis have been made on total samples and separately on the argillaceous fraction ($<$ 2 micron).

In total samples the content of quartz, feldspar, calcite and dolomite besides clay minerals has been determined. Here only the data referring to clay minerals are reported. As it is shown in figure 4 five fundamental clay mineralogic associations have been recognized in the whole examined territory:

A): illite-kaolin-chlorite, corresponding to northern Italy and the extreme south-western corner of the italian peninsula;

B): chlorite-illite-kaolin-smectite, corresponding to the north-eastern part of Italy;

C): smectite-chlorite-illite-kaolin, corresponding to central-eastern Italy;

D): kaolin-illite-chlorite, corresponding to central-western Italy;

E): kaolin-illite, corresponding to Sicily.

The more accurate data obtained from the fraction minor than 2 micron are reported in fig.5 which shows histograms related to representative samples.
As function of their approximative quantities the single individuated clay minerals may be classified as dominant, complementary and characterizing. These latter may allow to distinguish mineralogic associations with similar contents of dominant and complementary minerals. On the whole eleven mineralogic associations (A-M) may be distinguished (fig. 6). The A, B, C, D associations are fundamentally illitic-kaolinic; they show elevated quite variable contents of smectite and chlorite and may differ each from another because of the occurrence of interlayers as characterizing minerals. The E association decisively differs from the preceding ones because of the dominant character of smectite together with kaolin and illite. In the F association smectite is absolutely dominant. The G association is markedly composite. H and I associations are clearly kaolinic with also illite as associated dominant mineral. They differ among them for the occurrence of smectite in I association. L association is individuated by the occurrence of equivalent amounts of smectite and kaolin. M association is decisively kaolinic.

The above mentioned associations show a more clear regional distribution in the italian territory with regard to those obtained by total samples. They in fact correspond to clay mineralogic provinces partially originated, may be,by differential dispersal of the various clay species in sedimentary basins. The origin of the regional distribution of the various clay associations may be perhaps better ascribed to the lithological composition of the different parent rocks they derived from. Coarse relations at a regional scale between quality of the clay minerals filling sedimentary deposits and presumable parent rocks must be prudentially considered. On the basis of the available data the following statements might be drawn (fig. 7):

mineral	parent rock
kaolin	: clay, metamorphic rocks, acid eruptive rocks;
illite	: metamorphic rocks, arenaceous rocks;
smectite	: marlstone, clay, limestones (?);
chlorite	: basic eruptive rocks, marlstone (?);
interlayers	: marlstone, limestone (?).

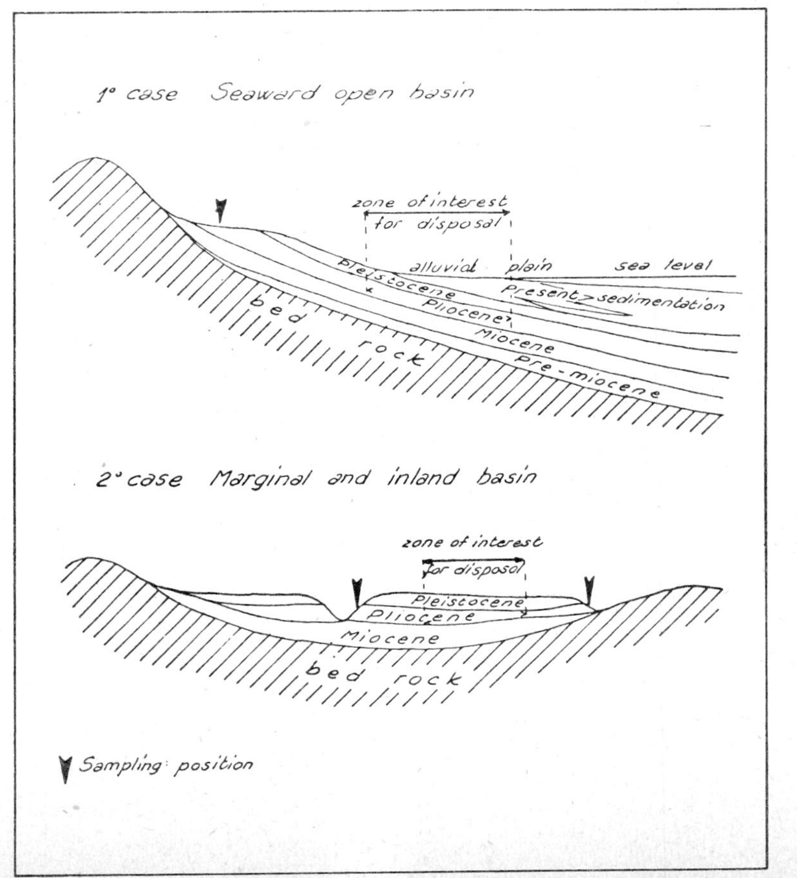

Fig. 2 Sampling position.

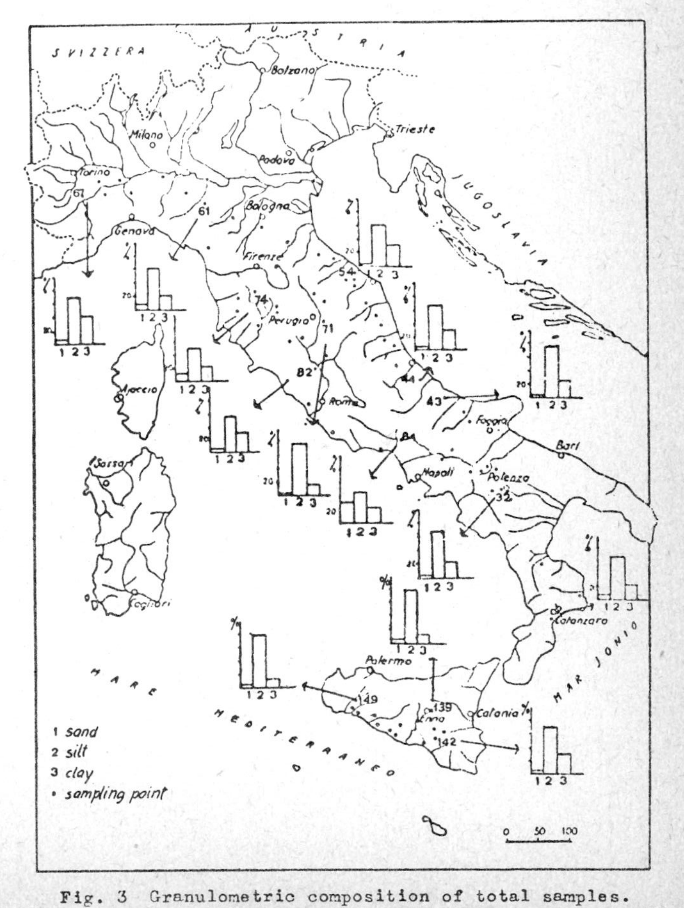

Fig. 3 Granulometric composition of total samples.

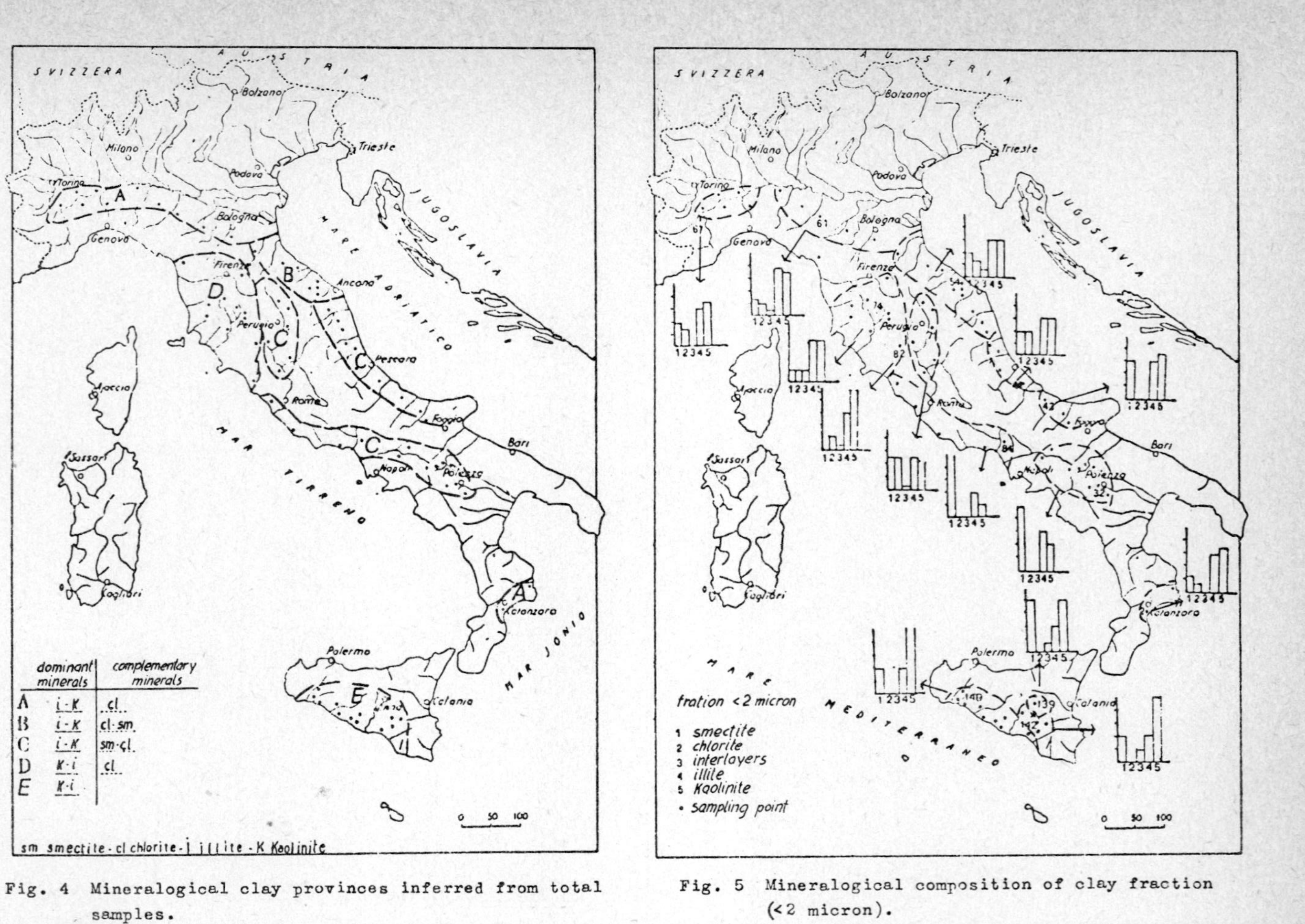

Fig. 4 Mineralogical clay provinces inferred from total samples.

Fig. 5 Mineralogical composition of clay fraction (<2 micron).

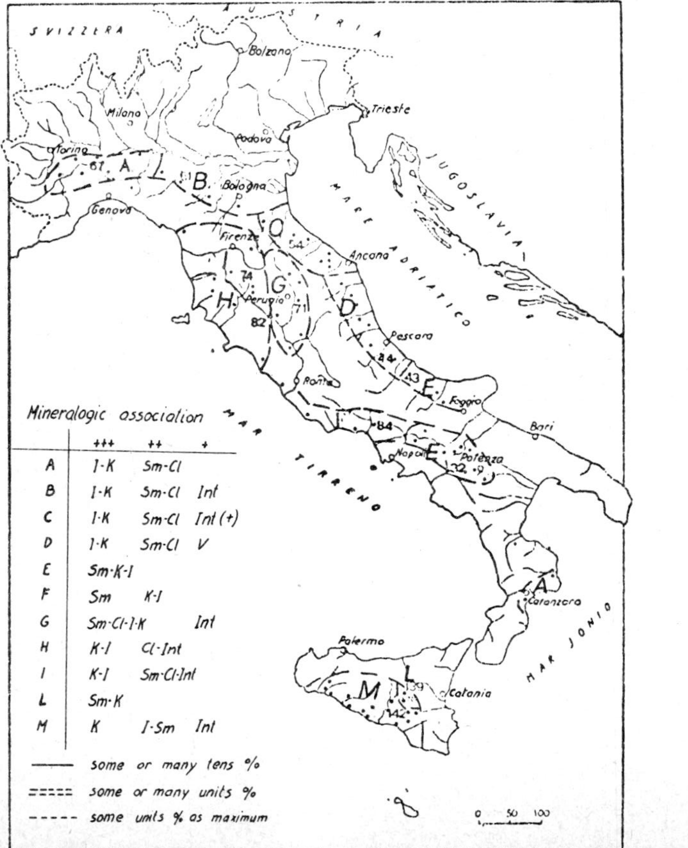

Fig. 6 Mineralogical clay provinces inferred from clay fraction (<2 micron).

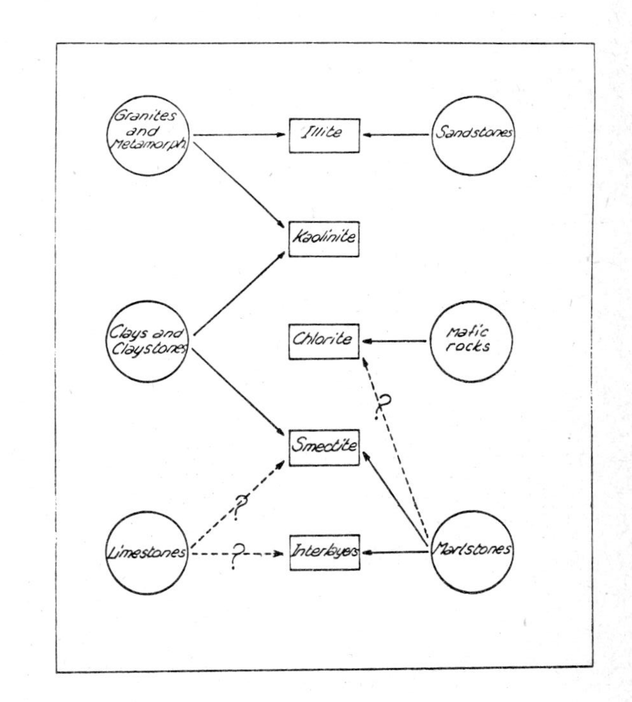

Fig. 7 Principal correlation between clay minerals and parent bedrocks as observed in the italian basins

The most evident facts arising from this study are:

1) Clay deposits correspond to well defined mineralogic provinces and show a clear regional distribution.

2) A coarse correlation seems to exist between lithological nature of ancient feeder basins and mineralogy of deep clay deposits.

The two above reported statements may be used on a probabilistic basis as a predictive tool for the inferring some mineralogic and structural characteristics of the deep deposits. They are not surely sufficient for an indipendent use in the selection of the most appropriate deposits but they are undoubtely usefull in orientating choice between situations resulted similar on the basis of other parameters. Deep deposits must indeed also correspond to the same mineralogic provinces ascertained by means of the study of their marginal outcropping parts. Parent rocks must have exercited in their regard the same rule recognized for outcropping clays.

Referring to the last statement above reported, the influence exercited by lithology of ancient feeder basin in composition and setting of derived sedimentary deposits may be better understood taking into consideration two opposite situations, for instance a sedimentary basin corresponding to a sandy erosional basin and a second basin corresponding to a clayey erosional basin (fig. 8). It is evident that the ratio sandy sediments/clayey sediments is much higher in the first case. This fact may be considered useful in orientating, still on a probabilistic basis, sitological choice directed towards clay bodies with poor or no sandy interbedding and linked operations.

The erosional stage of an ancient dominantly clayey series may strongly determine the dimension of the derived clay deposit. Figure 9 reports three cases of different development of clay bodies underlying an alluvial coastal plain according the erosional stage of a preceding clay formation having, as usually it is the case, important sandy component landward.It is evident that dimension and purity of the new clay deposit is greatest before the complete destruction of the marginal sands.

If some necessity exists with regard to the constancy of mineralogic composition of clay deposits it must be taken in mind that clayey sediments may display important variations from this standpoint. These variations may derived either from inheritance causes or from a selection operated by depositional mechanisms. Fig. 10 reports a scheme of a zonal distribution pattern starting from the coast as recognized in many part of the world. In a real case of present sedimentation studied in Italy (Golfo di La Spezia) the zonal series from the coast seaward in an interval of about twenty kilometers is resulted kaolin-illite-chlorite-interlayers. Some theoric speculations of the mentioned zonality may be drawn taking into consideration the variability of the position of coast line during the evolution of a sedimentary basin (figg. 11 and 12). The shifting of clay minerals belonging to the different zones is directed seaward during the regressive phase and landward in the ingressive phase. The geometry of the deposit builded up in a single complete transgressive cycle may be different according the initial ingressive or regressive character of the cycle. This fact may assume practical implication if a previously selected zone is an alluvial coastal plain. Whole clay bodies and single mineralogic species in the underground should be quantitatively more important if deposited in a cycle began with an ingressive phase. Generally speaking a sedimentary series is however made of a succession of complete transgressive cycles and its edges assume an S or saw profile (fig. 13). This implies that the probability of finding vertically continuous clay deposits under coastal alluvial plain is higher near the coast line.

As a general conclusion the results of the work done and the

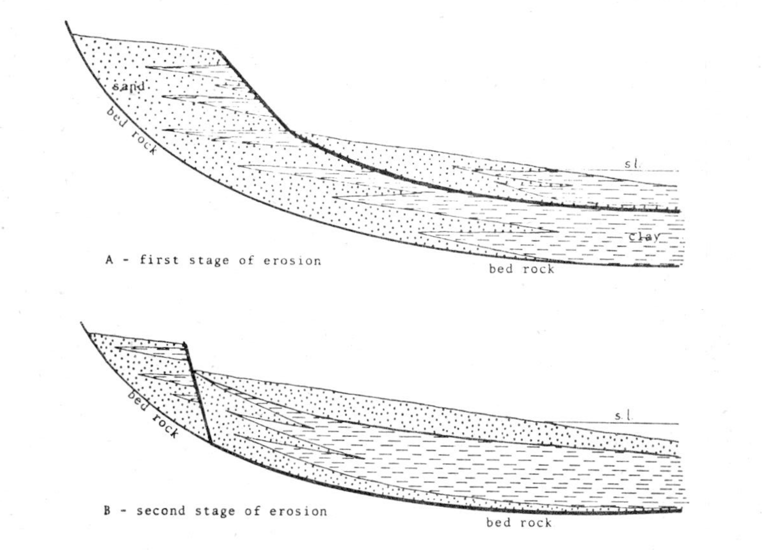

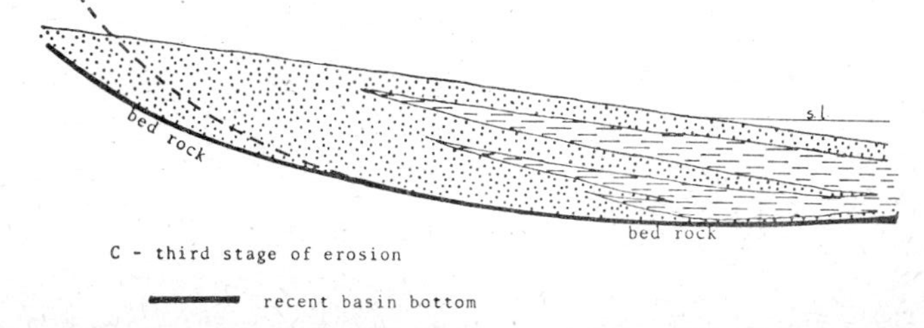

Fig. 9 Lithological variation in recent basin according to the different erosional stage of an older clay formation.

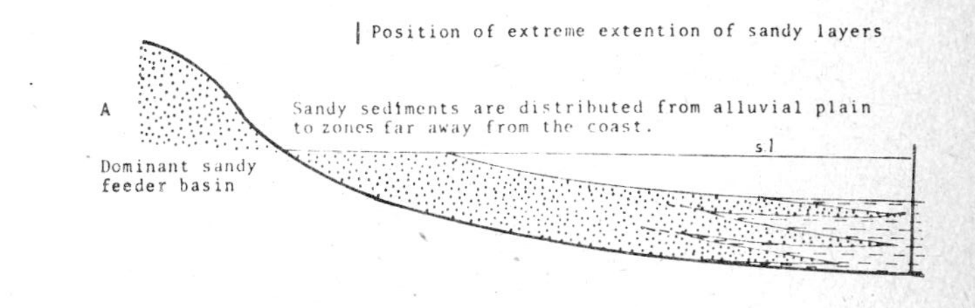

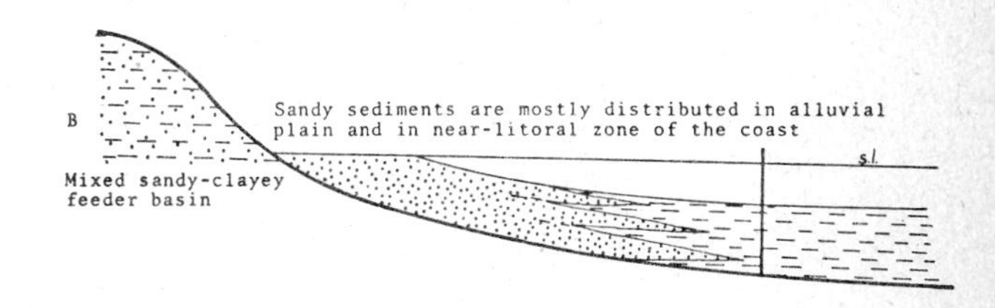

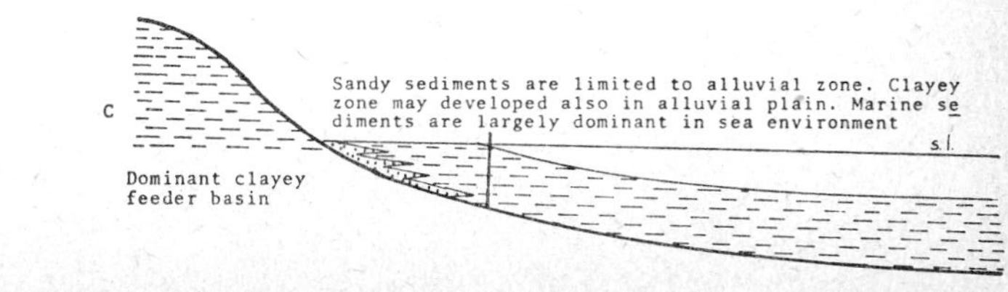

Fig. 8 Comparative extention of sandy and clayey sediments according lithological nature of feeder basins.

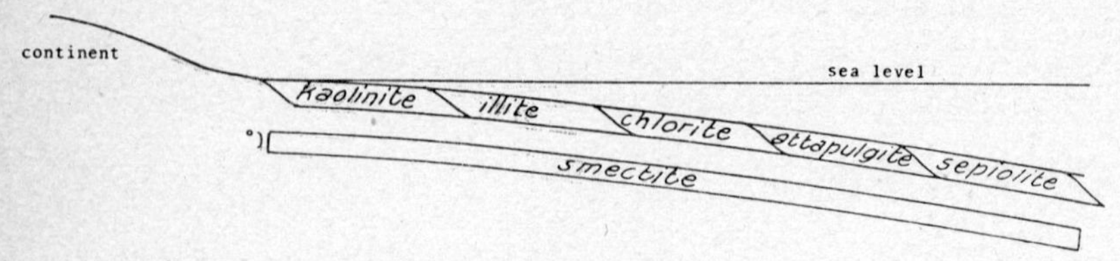

Fig.10 Scheme of lateral variations of clay types in a sedimentary unit from continent to deep sea.

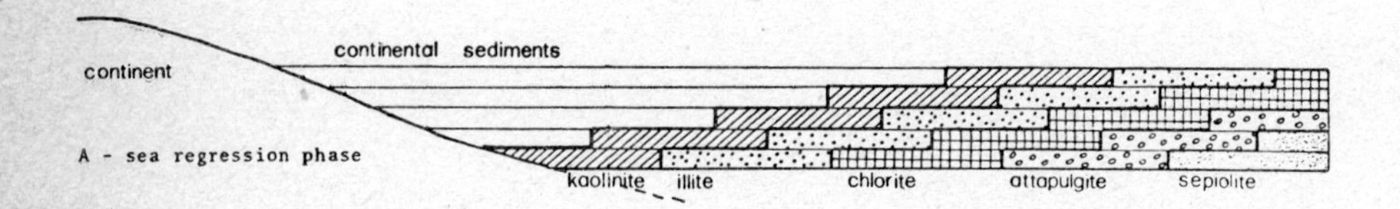

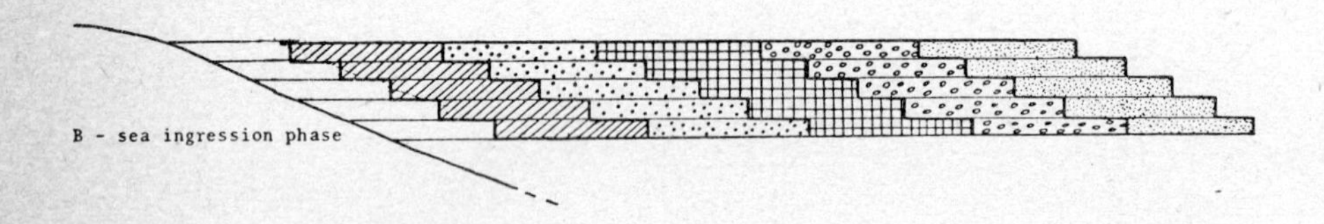

Fig.11 Schematic rapresentation of depositional setting of clay minerals under different conditions of sea transgression. Kaolinite,...,...,..., sepiolite series corresponds to deposition of clay minerals with sea depth. (An individual level corresponds to a synchronic sedimentary phase.)

considerations above reported allow the use of a naturalistic approach as an usefull tool in the predicting some characteristics of buried clay bodies. There is no doubt that the most relevant parameter qualifying a clay deposit is the size. From this point of view it can be affirmed that in some cases general informations are already available from other kinds of investigations elsewhere conducted. Furthermore non geologic parameter may also consitute selecting factors, as in the case of demography, infrastructure availability, local political and opposition problems and so own. Once the number of potential sites has been individuated, naturalistic considerations and superficial investigations allowing predictions on mineralogy, homogeneity, geometrical setting and other characters may orientate the choice both in the selection of the most suitable sites and in planning research operations in selected sites.

Acknowledgements:

The Commission of European Community for the permission of this communication.

Mr. Fernando Cevolani, CNEN, for his technical support.

REFERENCES

1 - COMMISSION OF THE EUROPEAN COMMUNITIES. 1979 - "European Catalogue of geological formations having favourable characteristics for the disposal of solidified high-level and/or long-lived radioactive waste".

2 - INTERNATIONAL ATOMIC ENERGY AGENCY. 1977 - "Site selection factors respositories of solid high-level and alpha-bearing wastes in geological formations". Wien. 1979.

3 - BRONDI A. 1979 - "Some considerations on a borehole in a clay formation". Proceedings of the NEA Workshops: "Use of argillaceous materials for the isolation of radioactive waste". Paris.

4 - PARHAM W.E. 1966 - "Lateral variations of clay mineral assemblages in modern and ancient sediments". Proc. Intern. Clay Conf. Jerusalem. 1, 135-145.

5 - QUACKERNAAT J. 1968 - "X-Ray analysis of clay minerals in some recent fluviatile sediments along the coast of Central Italy". Thesis. University of Amsterdam.

6 - VENIALE F., SOGGETTI F., PIGORINI B., DAL NEGRO A., ADAMI A. 1972 - "Clay mineralogy of bottom sediments in the Adriatic Sea". Proceeding International Clay Conference, Madrid.

7 - VENIALE F., SOGGETTI F., SANTAGOSTINO C. 1977 "La distribuzione dei minerali argillosi nei sedimenti di fondo del Mare Adriatico II "mesofossa" e "fossa" centro-meridionali" Geol. Appl. e Idrogeol. Bari Vol. XII, parte II.

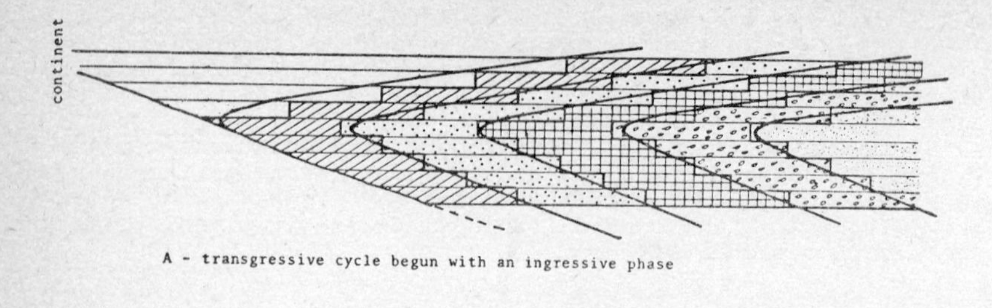

A - transgressive cycle begun with an ingressive phase

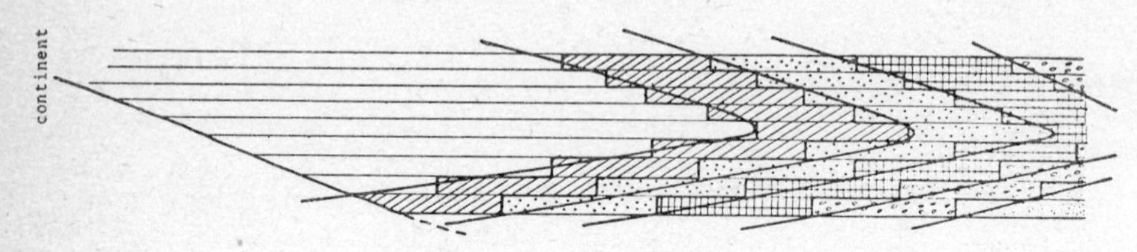

B - transgressive cycle begun with a regressive phase

Fig.12 Complete transgressive cycles.

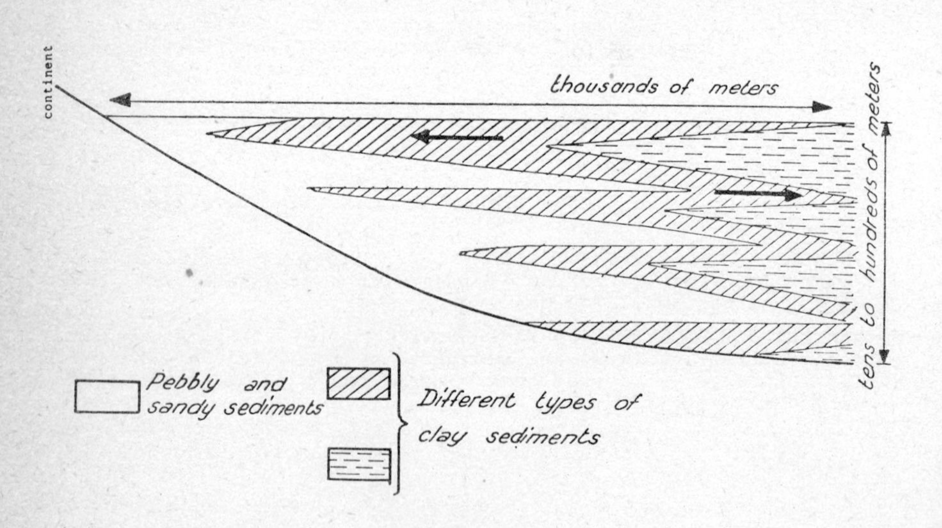

Fig.13 S - shaped setting and vertical zonal distribution of different clay sediments coastward.

DISCUSSION

M.F. THURY, Switzerland

Vous avez présenté une étude du genre recherche fondamentale concernant la composition minéralogique de différentes formations argileuses en Italie. La composition minéralogique n'est qu'un facteur parmi beaucoup d'autres qui influencent la sélection de sites. D'autres facteurs comme par exemple le régime hydrogéologique ou la sismicité ou des facteurs politiques sont très importants. Pensez-vous que la composition minéralogique va influencer la sélection des sites ?

A. BRONDI, Italy

Dans la phase des choix préliminaires la connaissance vérifiée ou présumée de la minéralogie des dépôts argileux profonds peut avoir un rôle seulement secondaire. Les autres paramètres géologiques, tels que la dimension des dépôts, la sismicité, la situation tectonique, en rapport avec d'autres paramètres non géologiques, tels que la démographie, le cadre politique et administratif général et local, sont sûrement plus importants que la minéralogie. Le rôle de la minéralogie peut au contraire devenir décisif pour le choix du site entre différents dépôts d'argiles qui semblent identiques lors de la sélection préliminaire.

En effet, compte tenu des différentes caractéristiques géotechniques présentées des différents matériaux argileux et sur la base des options adoptées pour l'élimination ou le confinement des déchets radioactifs, la minéralogie peut avoir une importance fondamentale.

Si le type d'option adoptée (élimination définitive ou provisoire) et la solution technique choisie (galeries ou forages profonds) sont définis a priori, la minéralogie peut également représenter un facteur de choix déterminant dans la phase d'orientation préliminaire.

Je voudrais néanmoins souligner que l'étude que j'ai présentée ne permet pas seulement l'acquisition de connaissances des éléments minéralogiques des argiles mais elle donne également une base, plus que simplement probabilistique, pour prédire jusqu'à un certain degré, les conditions d'homogénéité des dépôts argileux et les rapports volumétriques et latéraux avec les sédiments sableux auxquels ils peuvent être liés. Elle peut donc être utilisée soit comme facteur de sélection, soit pour l'identification et la localisation des conditions plus significatives dans l'étude d'un dépôt argileux profond et finalement comme une contribution à une programmation rationnelle des travaux à conduire.

SESSION III

Chairman - Président

L.J. ANDERSEN

(Denmark)

SEANCE III

RESULTS OF GEOLOGICAL INVESTIGATIONS OF THE MORS SALT DOME FOR THE DISPOSAL OF HIGH LEVEL RADIOACTIVE WASTE IN DENMARK

A.V. Joshi
ELSAM
DK-7000 Fredericia

ABSTRACT

Results of geophysical measurements, deep drillings, and hydrogeological testing are given. The salt in the preferred area for disposal is eminently well suited for radioactive waste disposal. The Cretaceous sediments above the salt have low permeability and good qualities of radionuclide sorption. The hydrology of the island of Mors is simple with low potential gradients. Mixing between the fresh waters in the aquifers near the surface with the underlying highly saline waters above the dome is retarded because of the large density differences between fresh and saline water. The regional potential gradients are too small to change this pattern. There are several indications that ultimately a stable interface boundary will develope. In the caprock and the Cretaceous transport today is by diffusion only.

1. Preliminary.

In March 1979 the board of ELKRAFT and ELSAM responsible for electricity production in Denmark accepted the Danish Government's request that they investigate the possibility of disposing high level radioactive waste in Denmark. Prelude to that was a previous investigation, phase 1, [1], which confirmed that it is possible to build a repository for radioactive waste in a suitable salt dome and that it is highly probable that a suitable dome can be found among the Danish domes.

In the present phase of investigations, termed phase 2, field investigations were therefore undertaken to find a salt dome with suitable geology.

These investigations are now complete, and the utilities expect to deliver next month, i.e. June, 1981, [2], their report to the Danish Government. It is the Government, who will evaluate these reports and decide on the adequacy of the Mors dome for radioactive waste disposal. Denmark does not have any nuclear power stations yet. The Government has made public acceptance - a referendum will be held for this purpose - a prequisite for the introduction of nuclear power in Denmark. A satisfactory solution to the waste disposal problem is considered by the Danish public to be a cardinal issue for Denmark going nuclear. Therefore it has been necessary at so early a stage (need for a disposal facility first arises in 2050) to find and investigate a suitable geological formation. The purpose, methods, and the time schedule of these investigations thus vary from those countries which are considering starting construction of waste disposal facilities in the next 5 to 10 years.

2. Geophysical Investigations.

The salt domes of Gørding, Linde, Vejrum, Sevel, Paarup, and Mors were investigated by seismic. In the selection of these domes emphasis was on their suitability also for a mined disposal facility. Relatively shallow salt domes were therefore chosen. The seismic investigations indicated that the shallower salt domes of Vejrum, Sevel, and Paarup might have undergone some leaching in the Quaternary times, and therefore the documentation of these domes for their suitability would require extensive hydrogeological investigations. The deeper domes of Gørding and Linde had the benefit of several hundred meters thick impermeable barrier of Tertiary clay sediment layers, while Mors had the simple hydrogeology of a small island location and showed no signs whatsoever of any leaching. The first drilling at Linde unfortunately missed the small dome, and as it lies deeper, and is smaller than previously expected, attention was focused on Mors.

Typical results of a seismic line can be seen in fig. 1. The Mors salt dome is one of the largest salt domes in Denmark. It lies beneath an island of that name and is situated nearly at the centre of the sedimentation bassin of North Jutland. Interpretation of the seismic surveys shows the dome to be elleptical in plan, having its major axis in the East West direction of 11.5 km, and minor axis in the North South direction of 8 km. The dome has an extensive overhang of about 2.5 km on the South Westerly flank (fig. 2). The top of the dome is at a depth of 600 m with a base at 5500 m. There is very little salt left at the base outside the dome proper. Future diapirism will thus be restricted through depletion of source material.

3. Internal Structure of the Salt Dome.

2 deep wells, 3.5 km deep, were drilled into the salt, fig. 3. The holes were drilled through Quaternary, Cretaceous, and caprock into Zechstein salt as shown in fig. 4. Erslev 1 was drilled through Na1 and Na2 layers. Salts of Veggerby potash zone K_2 were met between 2680 and 2930 m depths. Below, Na3 salt was encountered. Erslev 2 on the other hand was drilled almost entirely (735 to 3159 m) through Na1 (and perhaps Na2) salt [3]. Between 3159 to 3189 m Anhydrite layers were met, in an inverse bedding order (A1r/Ca2/A2). From 3189 to 3400 m the Na2 sequence was penetrated.

On the basis of core examinations, conventional borehole logs, and electromagnetic reflection logs the internal structure of the dome could be developed. The internal structure as envisaged by Prof. Richter-Bernburg [4] is depicted in fig. 2. From this figure it can be seen that Erslev 2 is drilled totally in Zechstein 1 and Zechstein 2 salt. The inverse bedding of the anhydrite layers met between 3159 to 3189 m in the borehole showed the salt sequence to be Zechstein 2 below 3189 m and Zechstein 1 above 3159 m. The entire section above 3159 m could be Zechstein 1, although because of its limited primary thickness in the Danish basin this is only possible with repeated foldings of Na1 layers. It is also likely that it is composed of both Na1 and Na2 layers (fig. 2) as these cannot be distinquished from each other from core examination. Judging by their lithological character the halite rocks of Zechstein 1 and Zechstein 2 are excellently suited for waste disposal. It is known that K and Mg salt series is totally absent in Zechstein 1 in the North Jutland bassin and is found in Zechstein 2 only at its interface with Zechstein 3.

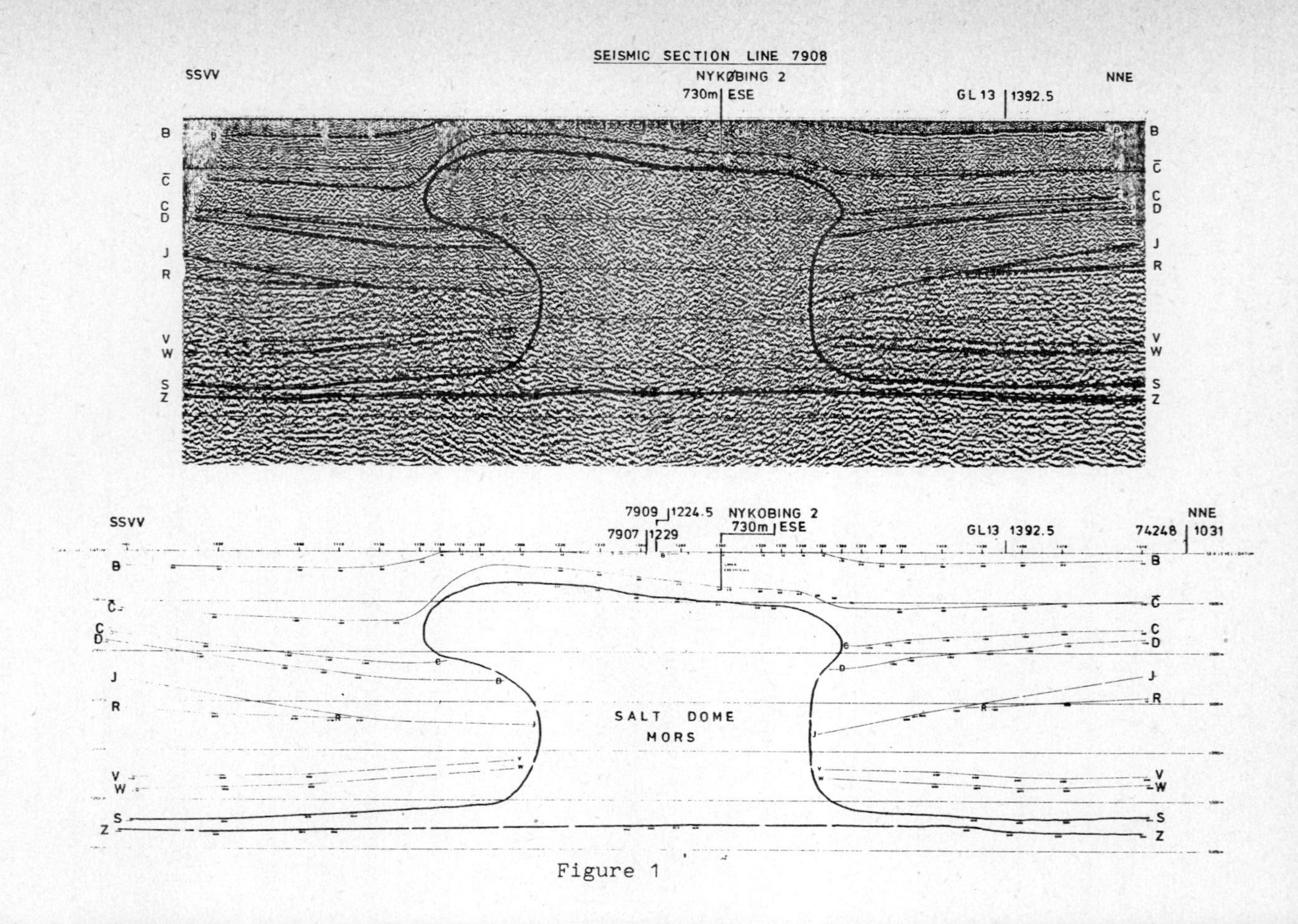

SEISMIC SECTION LINE 7908
SSVV
NYKØBING 2
730m ESE
GL 13 1392.5
NNE
B
C̄
C
D
J
R
V
W
S
Z
7909 1224.5
7907 1229
NYKOBING 2
730m ESE
GL13 1392.5
74248 1031
SALT DOME
MORS

Figure 1

MORS

LINE 7908

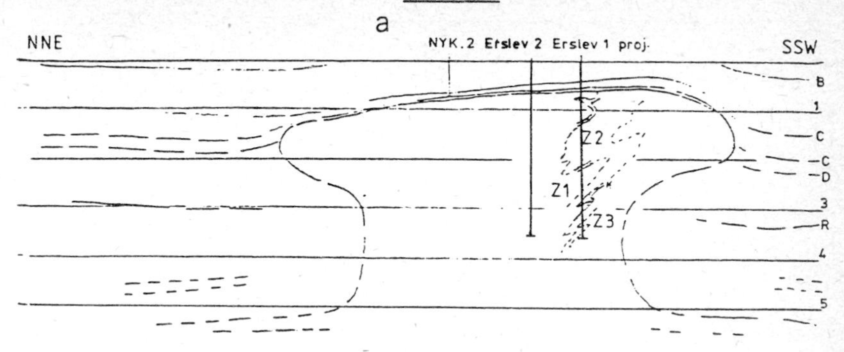

LINE 7907

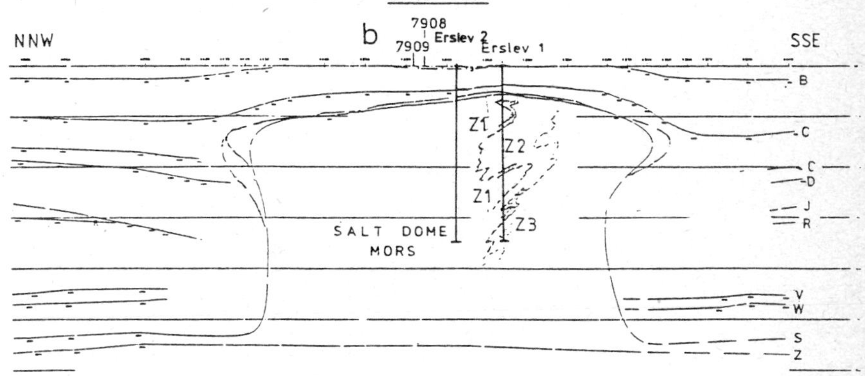

LINE 7909

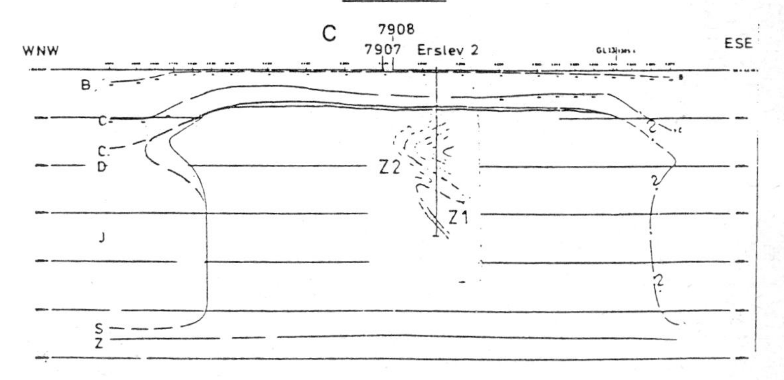

Based on Richter - Bernburg's interpretation (4)

Figure 2

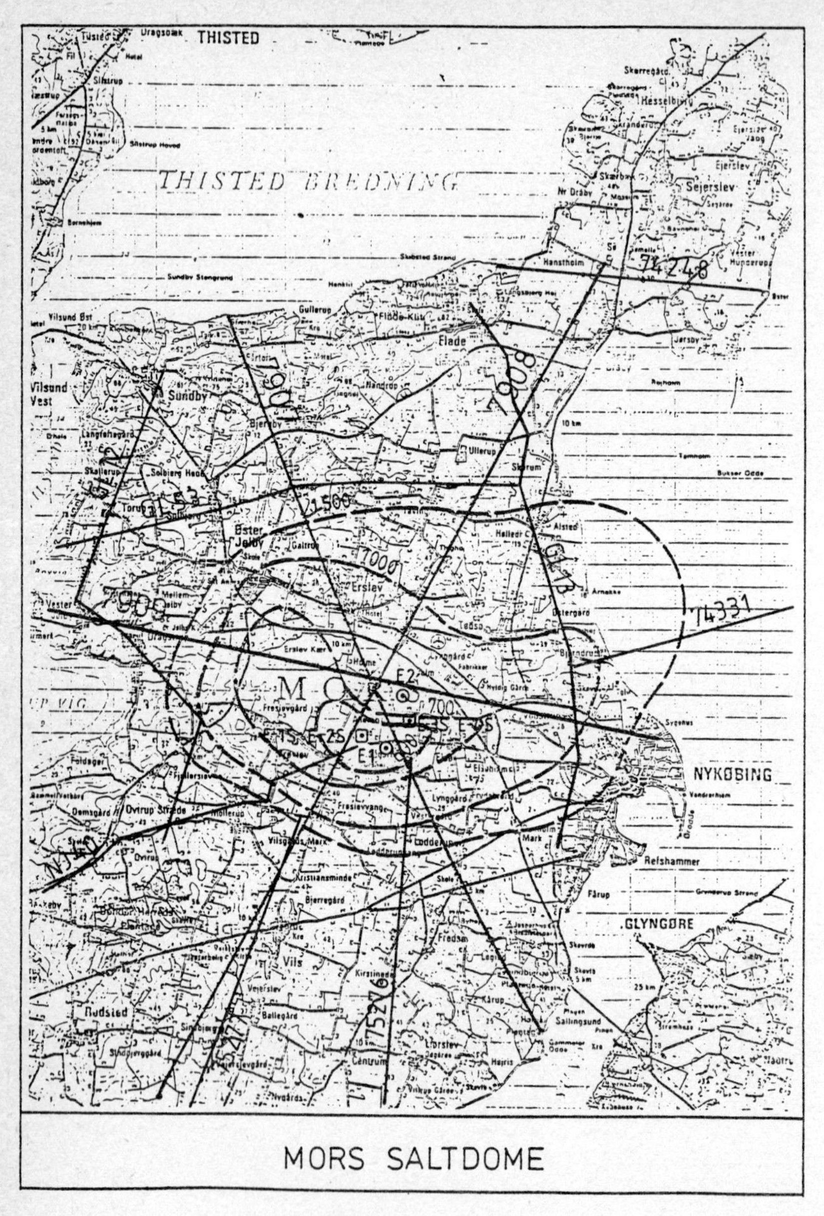

MORS SALTDOME

Figure 3

Comparison of Erslev 1 and Erslev 2
seen by Schlumberger-logs Gamma-Ray
and Sonic

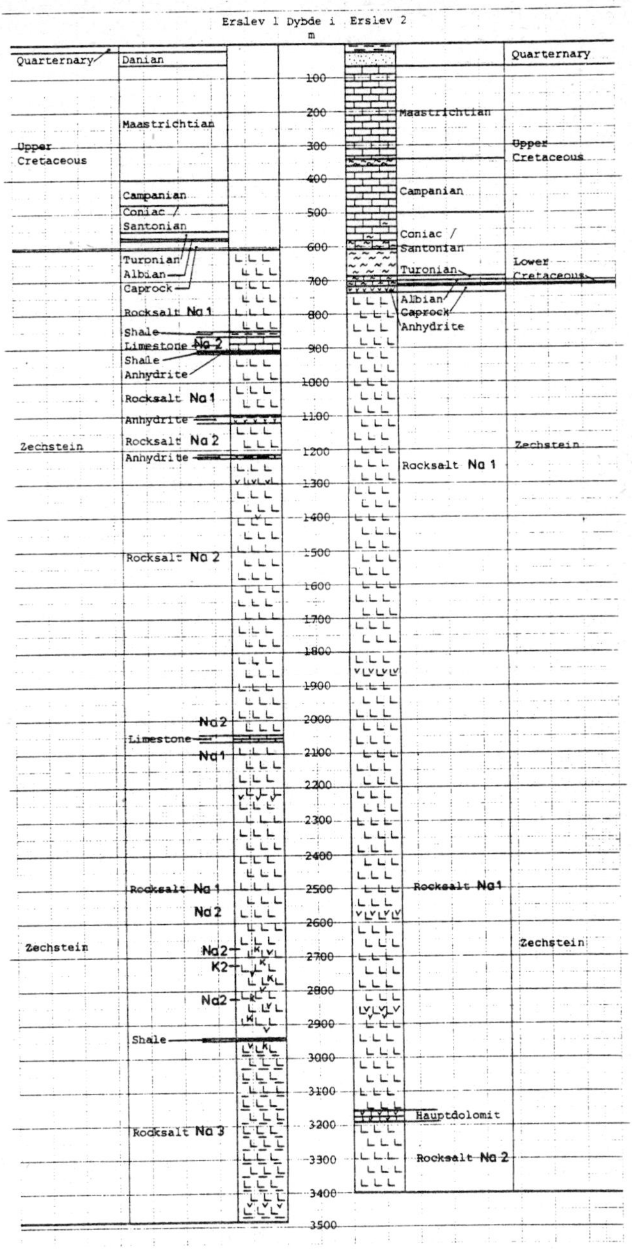

Erslev 1 Dybde i m Erslev 2

Quarternary
Danian
Maastrichtian
Upper
Cretaceous
Campanian
Coniac /
Santonian
Turonian
Albian
Caprock
Rocksalt Na 1
Shale
Limestone Na 2
Shale
Anhydrite
Rocksalt Na 1
Anhydrite
Zechstein
Rocksalt Na 2
Anhydrite
Rocksalt Na 2
Na2
Limestone
Na1
Rocksalt Na 1
Na 2
Zechstein
Na2
K2
Na2
Shale
Rocksalt Na 3

100
200
300
400
500
600
700
800
900
1000
1100
1200
1300
1400
1500
1600
1700
1800
1900
2000
2100
2200
2300
2400
2500
2600
2700
2800
2900
3000
3100
3200
3300
3400
3500

Quarternary
Maastrichtian
Upper
Cretaceous
Campanian
Coniac /
Santonian
Turonian
Lower
Cretaceous
Albian
Caprock
Anhydrite
Zechstein
Rocksalt Na 1
Rocksalt Na1
Zechstein
Hauptdolomit
Rocksalt Na 2

Figure 4

The section between Erslev 1 and Erslev 2 is likely to consist only of Zechstein 1 and Zechstein 2 salts. The Veggerby potash zone which crossed Erslev 1 at a depth of approx. 2700 m has a dip direction indicating that this potash zone is associated with the southern flank of the dome, and the base of the South West overhang.

The area around Erslev 2 (and between Erslev 1 and 2) is the preferred area for waste disposal. The anhydrite sections found there are probably anhydrite blocks floating in a Zechstein sea, particularly so at Erslev 2. Not being interconnected they cannot serve as pathways of water from the disposal area to a water bearing aquifer.

4. Salt Quality.

A physical, chemical, and mineralogical analysis has given the following results.

For Erslev 1 between 660 to 2500 m 65% of the tests showed the salt to be 98-100% NaCl and 0-2% $CaSO_4$, 30% showed 95-98% NaC1 and 1.5-5% $CaSO_4$, 5% of the tests showed 85-95% NaCl, 5-11% $CaSO_4$ and 0-4% insolubles. Potassium and Magnesium salts are about 0.01%. The Anhydrite showed up mostly in the form of clouds of small nodules.

The total water content in the salt of Erslev 1 in the same interval is less than 0.06%. In fact about 90% of the test specimens contained total water less than the measurable minimum value of 0.02%.

In Erslev 2, 99% of the salt above 3000 m consists of halite (NaCl) and anhydrite ($CaSO_4$). Potassium and Magnesium content is negligible.

The somewhat higher content of anhydrite containing some Gypsum at Erslev 2, in comparison to that at Erslev 1, gives a slightly higher total water content in the salt from Erslev 2, varying from 0 to 0.1% (with an isolated maximum of 0.22%). The mean water content of 77 samples is, however, less than 0.03%. In the disposal area the mean is only 0.017%.

The strength of the salt was very uniform - particularly at the shaft gallery level (about 1240 - 1300 m). The following results were obtained in uniaxial compression.

Erslev 1 σ_{cu} = - 24.3 MPa, V = 6.7% (31 tests)

Erslev 2 σ_{cu} = - 23.2 MPa, V = 9.6% (28 tests)
σ_{cu} = - 18.2 MPa, V = 14% (41 tests between 1300 m to 3000 m)

5. Borehole Convergence.

Borehole convergence was measured in Erslev 1, Erslev 2, and in a hole drilled in the Vejrum dome in 1973. Results from Erslev 1 are shown in figures 5 and 6. The convergence was calculated from caliper logs taken at various periods in a hole filled with oil mud with a specific gravity of 1.66. It can be seen from these curves that borehole convergence follows the expected pattern, i.e. decreasing with time. Convergence rates do not increase radically with depth. The accuracy of these field measurements undertaken with normal caliper tools is, however, limited. Laboratory experiments and calculations show that the convergence of the hole, even in case of an empty bore hole, will not give problems in the engineering design of the deep hole disposal concept. The holes Erslev 1, 2, and Vejrum are still kept open, and we hope to continue these convergence measurements perhaps with international cooperation.

6. Hydrogeological Investigations.

4 holes were drilled to a depth of 550 m - which is almost to the caprock level. The holes were placed in 2 nests about 1 km apart. Distance between the holes in a nest was approx. 30 m.

Over 60 pumping tests were performed. Injection tests and labelled slug tests were also performed. Cores, logs, and impression packer tests gave a good idea of the formation [5].

Fig. 7 shows an arrangement of a straddle packer test, fig. 8 and 9, results of permeability and porosity measurements.

The formations above the salt are shown in fig. 4. At Erslev 2 they consist of Quaternary (67 m), Maastrichtian (272 m), Campanian (275 m), Santonian (83 m), Turonian + Albian (13 m), Lower Cretaceous (5 m), and Caprock (20 m). The hydrogeological wells 1S and 2S ended about 40 m into the Santonian (approx. less than 60 m above caprock level), whereas 3S and 4S ended about 92 m in the Campanian.

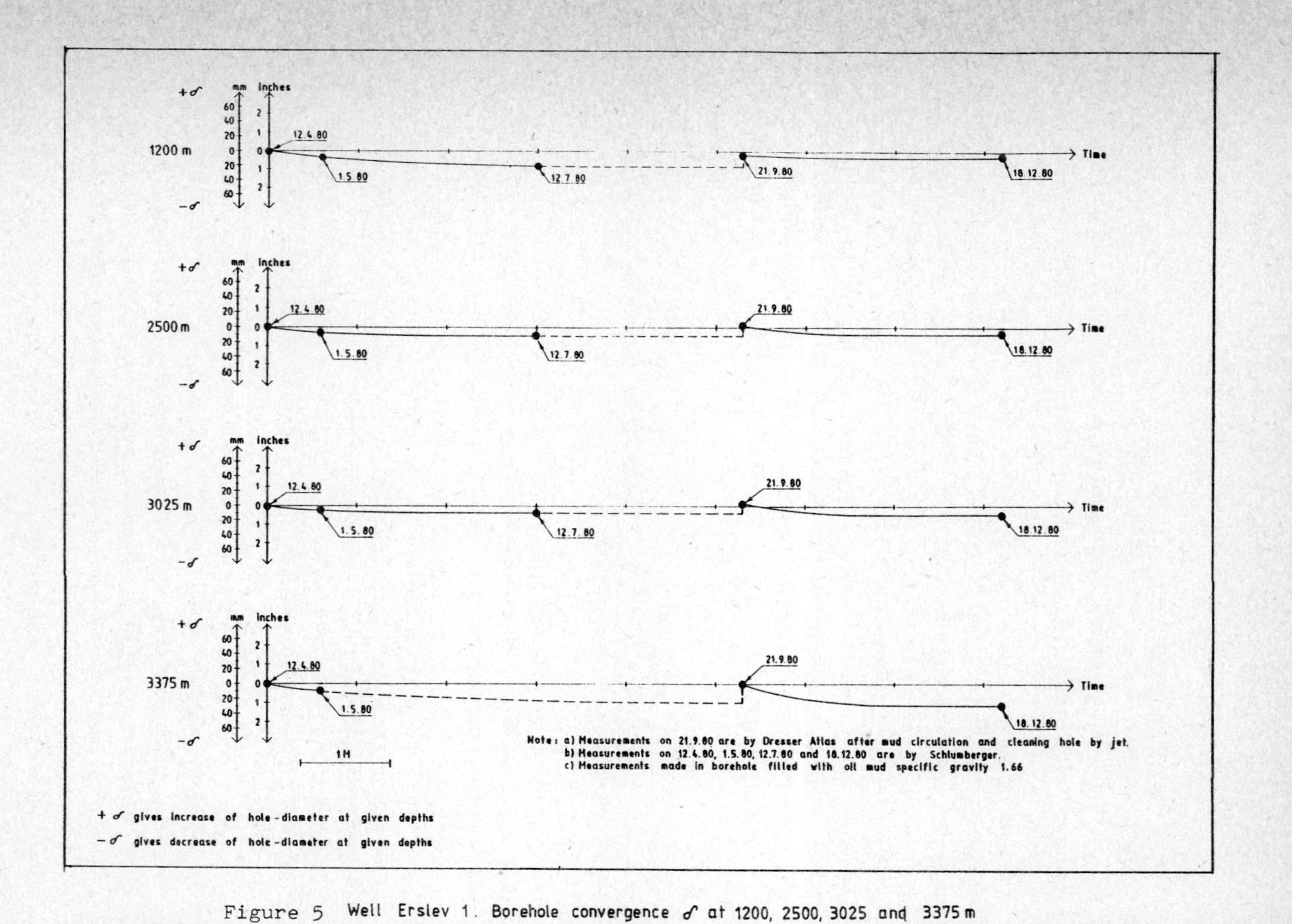

Figure 5 Well Erslev 1. Borehole convergence δ at 1200, 2500, 3025 and 3375 m

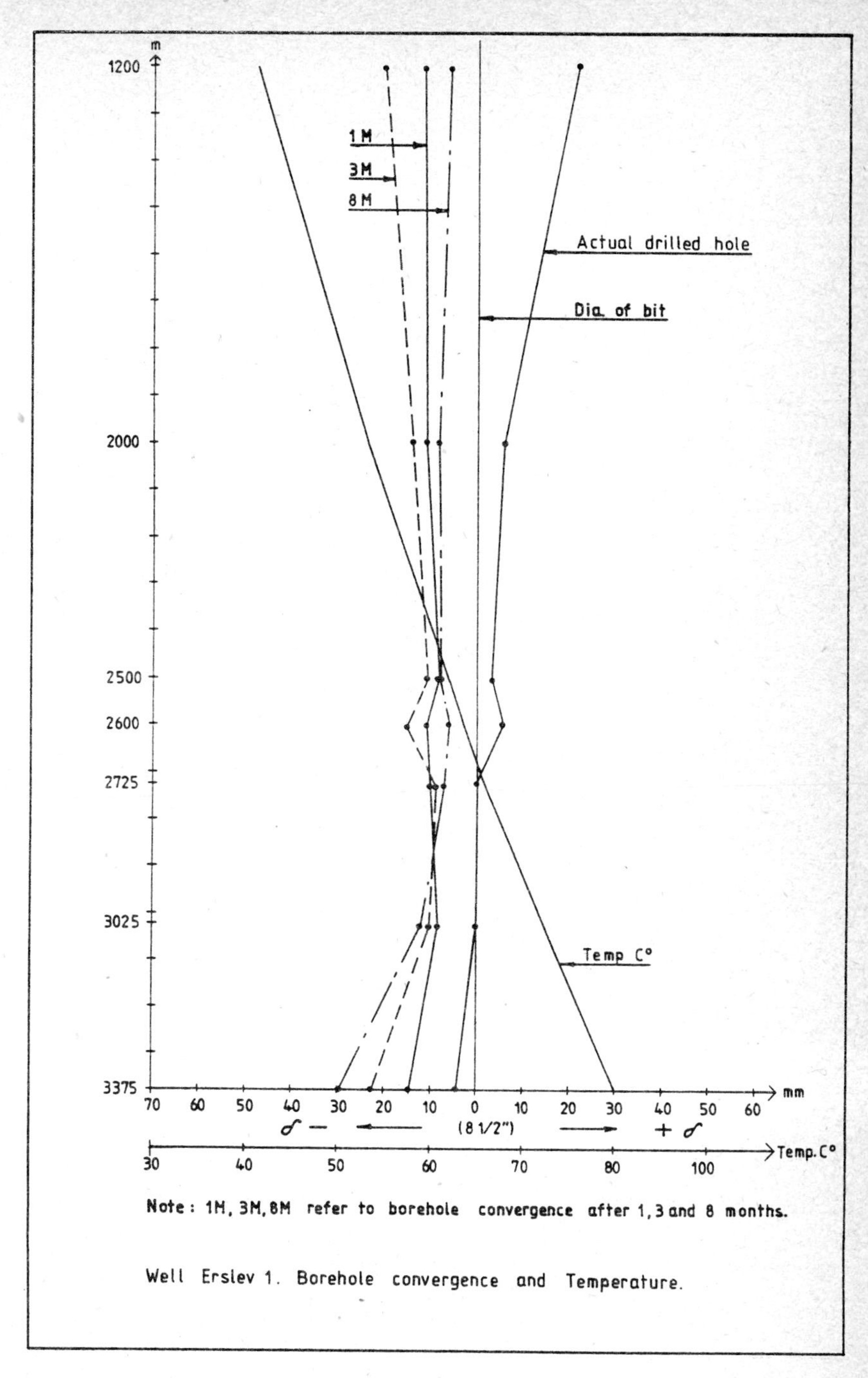

Note: 1M, 3M, 8M refer to borehole convergence after 1, 3 and 8 months.

Well Erslev 1. Borehole convergence and Temperature.

Figure 6

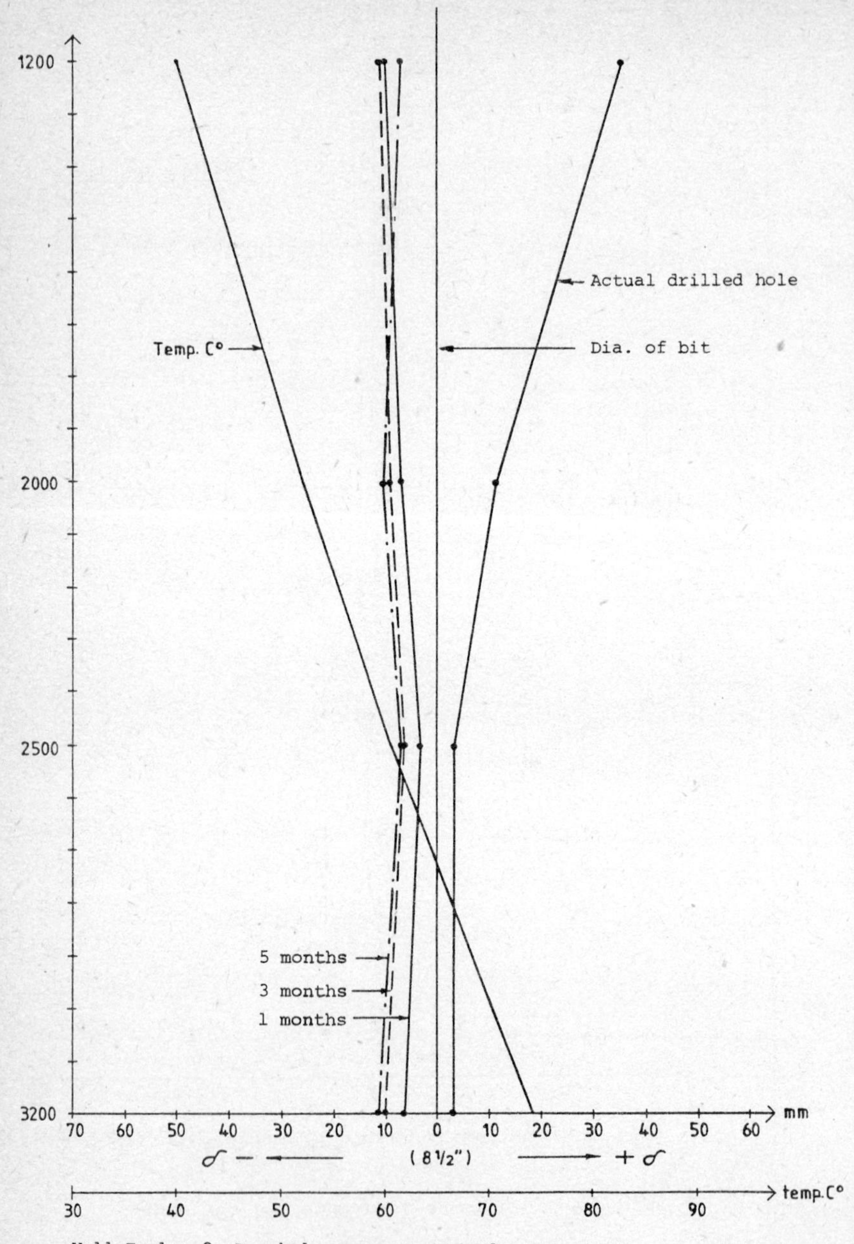

Well Erslev 2. Borehole convergence and temperature.

Figure 6 A

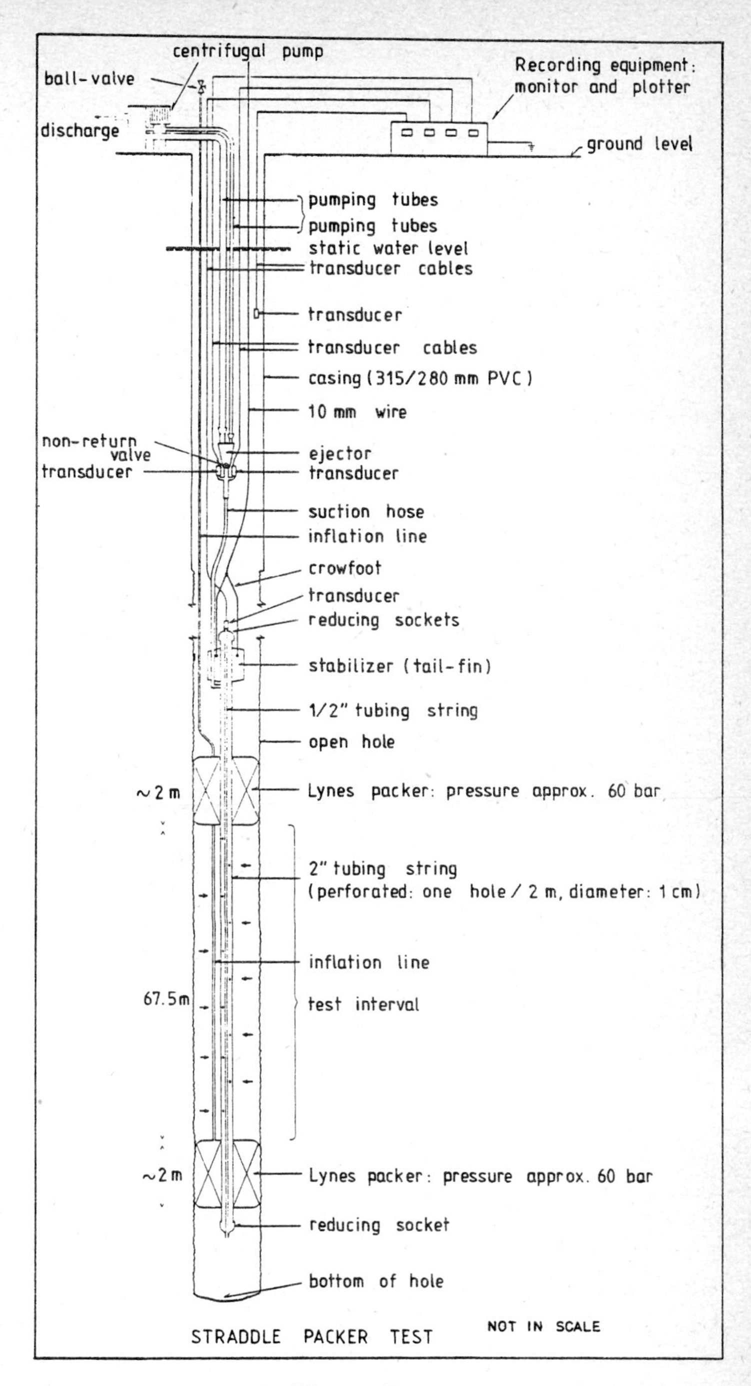
centrifugal pump
ball-valve
Recording equipment: monitor and plotter
discharge
ground level
pumping tubes
pumping tubes
static water level
transducer cables
transducer
transducer cables
casing (315/280 mm PVC)
10 mm wire
non-return valve
transducer
ejector
transducer
suction hose
inflation line
crowfoot
transducer
reducing sockets
stabilizer (tail-fin)
1/2" tubing string
open hole
~2 m
Lynes packer: pressure approx. 60 bar
2" tubing string
(perforated: one hole / 2 m, diameter: 1 cm)
inflation line
67.5m
test interval
~2 m
Lynes packer: pressure approx. 60 bar
reducing socket
bottom of hole
STRADDLE PACKER TEST
NOT IN SCALE

Figure 7

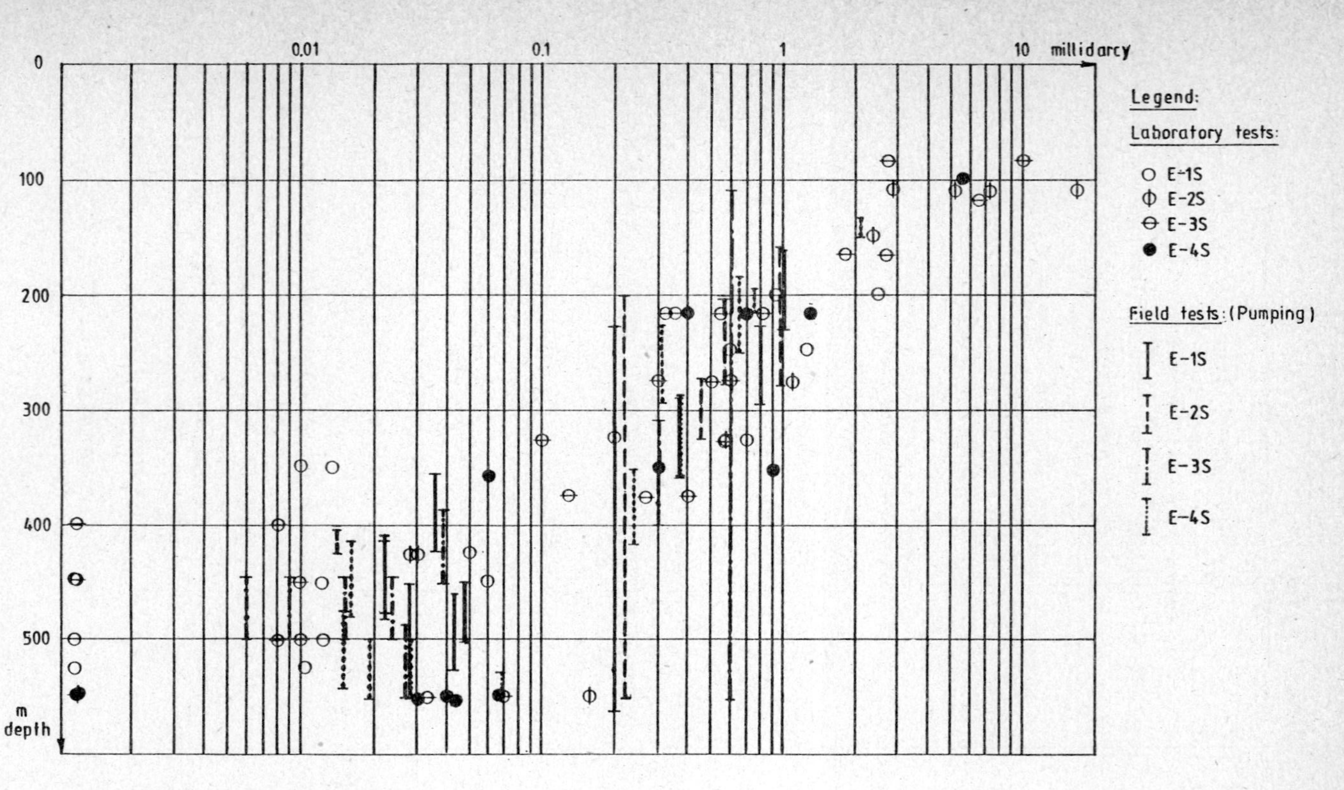

Hydraulic conductivity E-1S, E-2S, E-3S, E-4S
(It is assumed that 1 millidarcy is equivalent to 1.0×10^{-8} m/s)

Figure 8

A drill stem test in Erslev 2 proved the caprock there to be impermeable. The caprock has been investigated by Fracture Identification log, and fractures in the caprock have been detected only in the topmost 2 m, where the porosity is 1.5%. The caprock is dense with a porosity from FDC log of 0%. The seismic results show no sign of any leaching of salt.

Porosities were measured on cores by different methods and at different laboratories, as well as deduced from electrical logs (fig. 9). In the Maastrichian a porosity of the order of 40% was obtained. The porosity decreased abruptly in the Campanian, where porosities decreasing with depth from about 30% at 350 m, to about 20% at 550 m were obtained. Below 550 m and in the Lower Cretaceous porosities decrease from 25 to 10% (on the basis of logs from Erslev 2).

Permeability was determined on the basis of pumping tests, injection tests, and laboratory measurements. Results from field tests and core measurements at different laboratories and by different methods all show good agreement.

Permeabilities are of the order of 1 millidarcy (10^{-8} m/s) in the Maastrichtian and 0.01 millidarcy at 550 m depth in the Campanian (fig. 8). The caprock is impermeable.

Permeabilities decrease with depth in all 4 wells. Considering the distance between the wells and the different methods used for measurements in the field and laboratory, the agreement in the results is remarkable. Pumping tests give matrix plus fissure permeability, while the laboratory measurements give matrix permeability. This would indicate that although a number of fissures were detected from cores, fracture identification logs, and impression packer tests, the fissures do not yield any water. The holes were reamed and brushed with clean water to remove problems of mud cake. The mud used for drilling was fresh water with an admixture of Antisol B *) with a specific gravity between 1.00 to 1.04 - i.e. considerably less than the sp. g. of saline formation waters. The static pressure of the drilling fluid in the Campanian was thus less than the hydrostatic pressure in the saline formation waters - so mud intrusion during drilling was probably not so important a problem at these depths. It is important here also to remember that yield of the wells was very small, and the permeabilities were determined from recovery tests.

---

*) (Sodium carboxymethylcellulose - soluble in water-ideal formula ($C_{30} H_{43} O_{26} Na_3$)N))

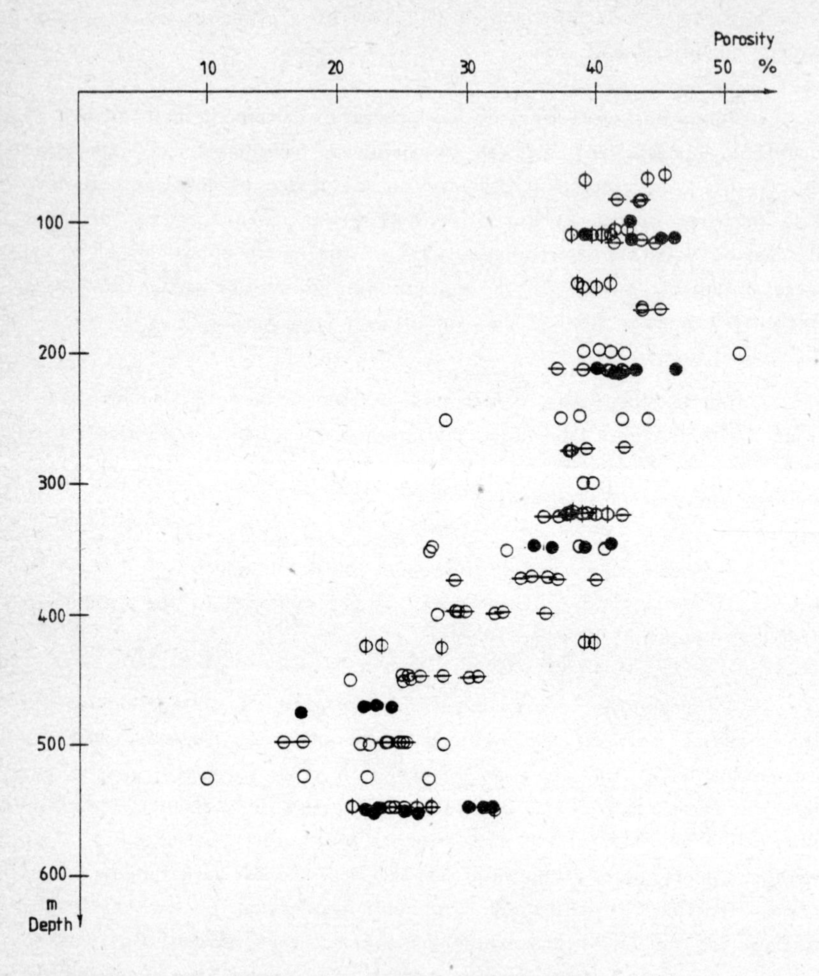

Porosity
10
20
30
40
50
%
100
200
300
400
500
600
m
Depth
WELL E-1S, E-2S, E-3S AND E-4S, POROSITY VERSUS DEPTH
E-1S
E-2S
E-3S
E-4S

Figure 9

The only plausible reason for the low yield from these wells, although fissures were met during drilling, is that at these depths they are sealed by sedimentation and geochemical processes (diagenesis). Geologists from the Danish

Geological Survey, however, prefer the explanation that the central area of Mors - encompassing Erslev 1 and 2, and the 4 hydrogeological wells - is an atypical compression zone [6], and the fissures are therefore closed. I do not quite agree with that idea. The low permeabilities obtained in the Cretaceous at these depths are perhaps the rule for these sediments rather than an exception. One should not compare these permeabilities with those in oil bearing stratums as the geochemistry is quite different.

The salinity of the formation waters was determined from logs, core measurements, and pumping tests. In the Campanian the pumping tests yielded only small amounts of formation water. As this was mixed with the drilling fluid, formation salinity had to be interpolated. However, when all these results are interpretated a salinity of about 140 g/l NaCl (45% saturation) is obtained at 550 m in S4.

Retention measurements were made on selected cores in the Upper Cretaceous. The results from all the 4 wells are strikingly similar [7].

The retention factor is defined as $R_f = \frac{t_o}{t_r}$

where t_o is the travel time of the nuclide when there is no retention, while t_r is the corresponding time taking retention into account.

Experiments were performed corresponding to bulk sodium chloride concentrations of 1M and 5M for the equilibrating solution respectively. Results from batch-type experiments [7] are as follows:

Mean retention factors and standard deviations.

NaCl	1M	5M
R_f (E_u)	$(5.5 \pm 3.8) \cdot 10^{-5}$	$(5.2 \pm 4.4) \cdot 10^{-5}$
R_f (C_o)	$(7.8 \pm 4.4) \cdot 10^{-3}$	$(9.0 \pm 4.1) \cdot 10^{-3}$
R_f (S_r)	$(8.9 \pm 3.1) \cdot 10^{-2}$	$(9.8 \pm 3.6) \cdot 10^{-2}$
R_f (C_s)	$(1.4 \pm 1.3) \cdot 10^{-1}$	$(2.1 \pm 1.3) \cdot 10^{-1}$

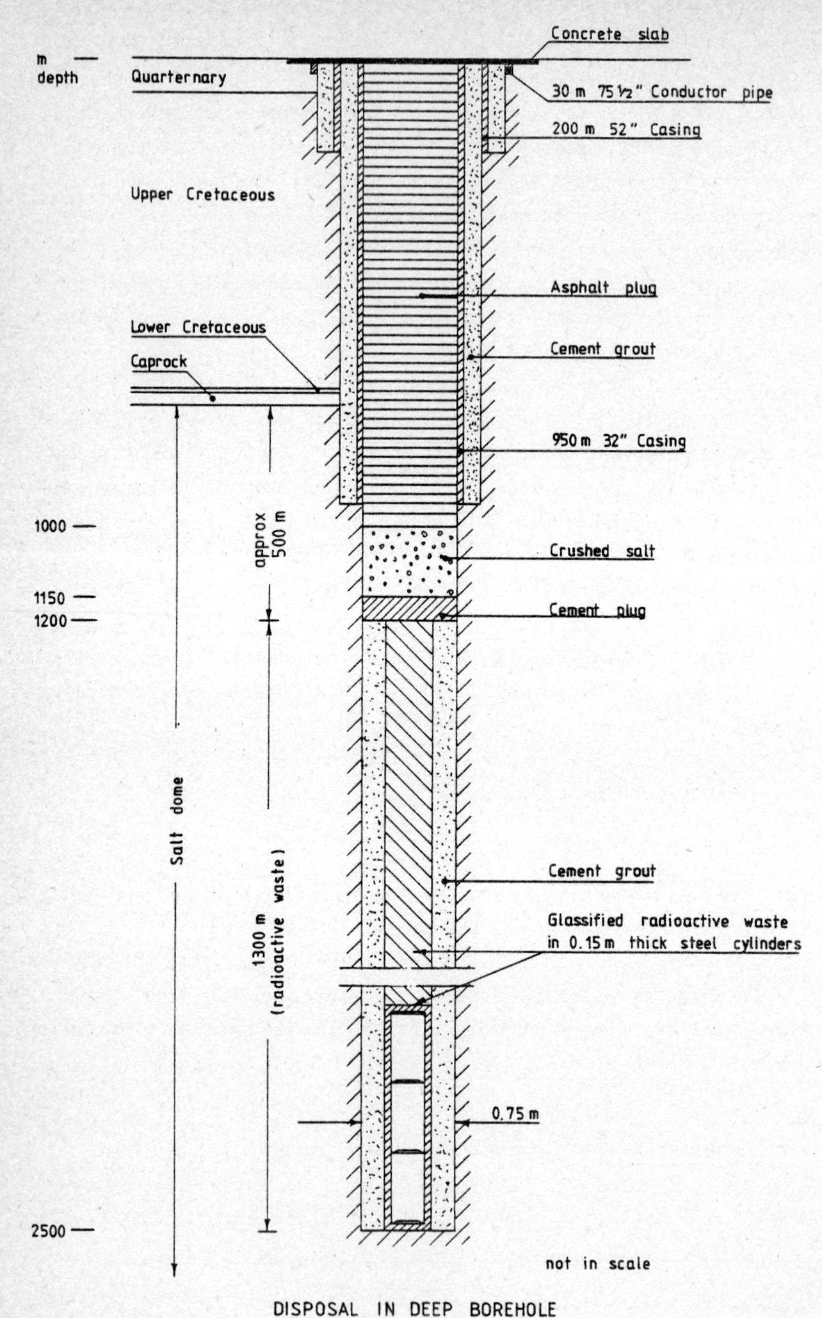

m depth
Quarternary
Upper Cretaceous
Lower Cretaceous
Caprock
1000
1150
1200
2500
approx 500 m
Salt dome
1300 m (radioactive waste)
Concrete slab
30 m 75½" Conductor pipe
200 m 52" Casing
Asphalt plug
Cement grout
950 m 32" Casing
Crushed salt
Cement plug
Cement grout
Glassified radioactive waste in 0.15 m thick steel cylinders
0.75 m
not in scale
DISPOSAL IN DEEP BOREHOLE

Figure 10

For salt concentrations below 0.5M the retention factors decreased dramatically and could not indeed be measured.

The experiments showed that the Upper Cretaceous as found at Mors exhibits retarding effect on cationic species as Cs^+, Sr^{2+}, Co^{2+}, and Eu^{3+}, whereas anionic species as Cl^- and TcO_4^- were found to move with the water front, i.e. without retardation. Even in cases of moderate water flows through the chalk formation, migration of actinide cations, as Am^{3+} and Pu^{3+} effectively will be retarded.

7. Deep Hole Disposal.

While disposal in the Mors salt dome is possible using both a mined shaft gallery system or disposal in large diameter holes drilled from the surface [2], the author prefers the latter for the disposal of the high level radioactive waste in Denmark. Fig. 10 shows such a disposal method and is self explanatory.

8. Implications of the Investigations on the Suitability of Mors as a Suitable Geological Formation.

Although all field and laboratory investigations in the present phase of the project are now complete - the evaluation of the results and their synthesis is still under way. Therefore the author has put his own views forward in this section. This is mainly to attract criticism at this meeting of experts from many countries and thereby help in the synthesis of the results obtained.

The radioactive waste disposal problem is an interdisciplinary problem. The geological formation cannot be considered on its own, but must also be considered in connection with the engineering design of the disposal facility. Future natural events as well as man-made intrusions must be considered with a balanced viewpoint. After all, radioactive waste in its toxicity is similar to other toxic products we have learned to live with [8 to 12]. J.J. Cohen writes very lucidly about the toxicity of nuclear wastes and asks the question "Why is nuclear waste such a special problem? Is actual technological risk the cause for concern, or is it really the public's perception of that risk? If the latter is the case, then would a prolonged and costly research programme really tend to relieve the fears or might it have the opposite effect?" [12].

Considering specifically Mors and the deep hole disposal concept the following points are worth mentioning.

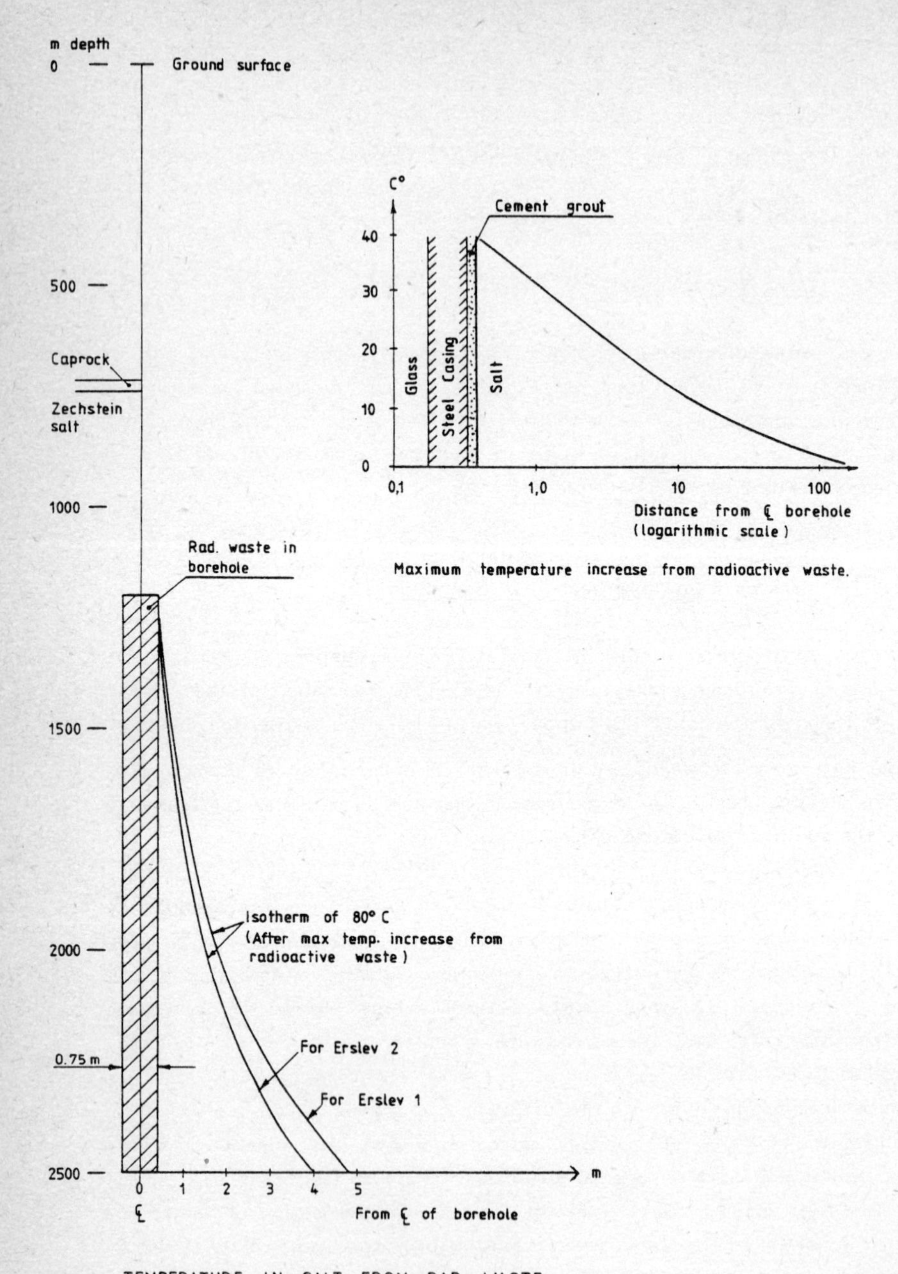

TEMPERATURE IN SALT FROM RAD. WASTE.

Figure 11

8.1. The Salt Barrier.

The primary barrier is salt. Engineering design including the encapsulation of the glass in a 15 cm thick steel cylinder and a minimum 40 year cooling time ensures:

a) Low temperatures in the salt-steel interface (fig. 11).

b) Even if large quantities of Carnallite were found 3.5 m away from the borehole, the temperature at 2500 m depth after taking into acccount temperature increase from radioactive waste ($60^oC + 20^oC = 80^oC$) will not release crystal water from the Carnallite. Higher up the Carnallite could be quite adjacent to the walls of the borehole (fig. 11) and still not release water. There is some uncertainty regarding the temperature, at which Carnallite releases crystal water. Normally 110^oC is the accepted figure, but Herrmann quotes that at temperatures of 80-85^oC some water is released [13]

c) The total water content in the salt is very small, and even pessimistic calculations of brine migration give insignificant amounts of brine - not at all enough to corrode even a small hole in the cylinder.

d) The homogeneous nature of the Na1 and Na2 salt series found in Erslev 2 proves, that the presence of significant amounts of K and Mg salts in the vicinity of the hole is unlikely.

e) Anhydrite layers, which may be found in the neighbourhood of Erslev 2 and at the depths contemplated for radioactive waste disposal, will not be continuous, but only in the form of blocks of limited lengths. They cannot therefore form a passage to a water bearing aquifer.

f) The average diapirism (fig. 12) is insignificant. It is at present on an average 0.005 mm/year over the last several million years. The depletion of the salt supply to the base of the Mors dome will result in a steady decrease of the average rate of diapirism in future geological times. The heat content in the waste is too small to influence these rates, and calculations showed only minute increases in the rate of surface uplift by diapirism. The total upheaval of the soil surface above the dome due to thermal expansion caused by heat from the waste is only 6-8 cm and occurs after a few thousand years.

Period	Age Millions of Years
1 Quaternary	2
2 Tertiary	65
3 Cretaceous	136
4 Jurassic	190
5 Triassic	225
6 Permian	280

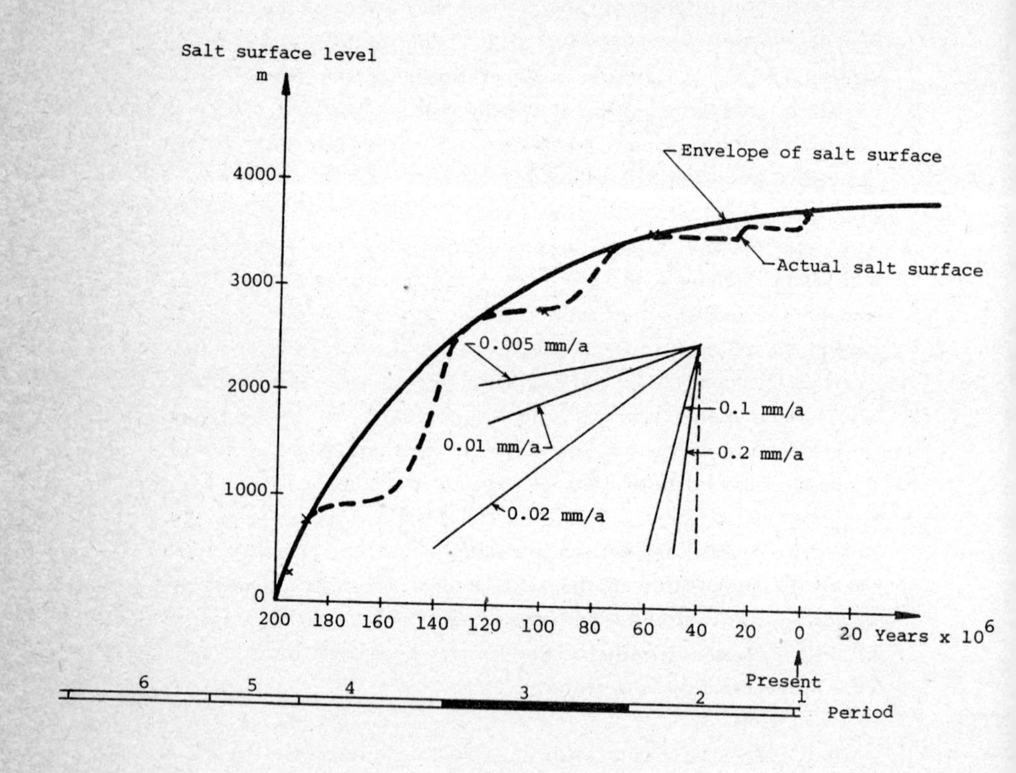

DIAPIRISM OF MORS SALTDOME.

Figure 12

g) The volume of salt necessary for waste disposal - including a 200 m safety barrier - is less than 2 km^3. The Mors dome with a salt volume of about 230 km^3, at depths suitable for disposal, provides a very substantial safety margin.

8.2. The Hydrogeological Barrier.

a) It must always be remembered that it is only a secondary geological barrier. The salt itself is the more important and primary barrier.

b) The low and uniform values of permeabilities observed, when considered together with the very low potential gradients obtained on the island location of Mors, give extremely slow horizontal hydraulic velocities below 550 m depth ($v < 0.02$ mm/year). The vertical velocities in this region are of the order of 0.002 mm/year.

c) Solgaard and Skytte Jensen [14] suggest that variation of pore water composition with depth follows a trend which indicates that diffusion of salt from the top of the dome is the only migration mechanism. I believe the salinity in the formation waters is a result of transportation of pore waters through consolidation, in times far gone, of the layers immediately above the dome. The pressure increase over millions of years through sedimentation of overlying layers causing this very slow transportation of saline porewater [2]. This process is now in equilibrium, and today only diffusion prevails. The observed decrease in Magnesium content at larger depths is suggested to be due to dolomitisation reactions [14].

d) The large density difference in the fresh and saline waters gives rise to an interface between the fresh water and saline water. Significant flows across this interface boundary - considering the conditions existing at Mors - are unlikely. The regional potential gradients are too small to change this pattern. Ultimately a stable interface boundary will develope where mixing across the interface boundary will largely only be by dispersion and diffusion (fig. 13). That this stable boundary has not been reached yet after millions of years, is but another indication of the impermeability of the underlying Cretaceous strata.

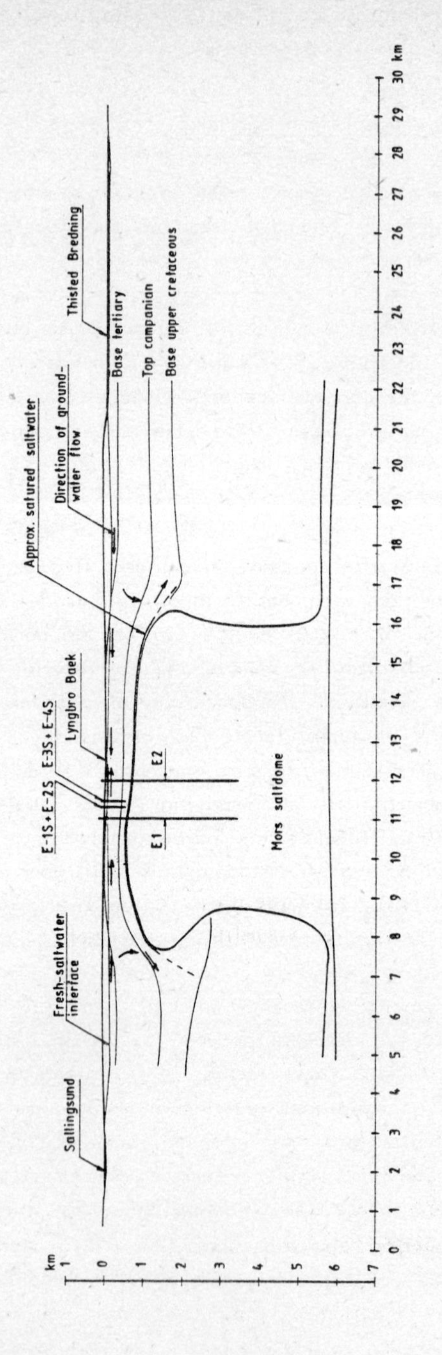

Figure 13 Hydrological model

e) Any release of radioactivity above the caprock - by an unlikely and hypothetical reason - will migrate to the biosphere only after several million years, when radioactivity has decayed to be of no consequence whatsoever. Diffusion is the main transport mechanism for migration in the lower parts of the Cretaceous and caprock. Near the saltwater fresh water boundary the process will be of diffusion across the boundary with dispersion in the fresh water itself.

The Cretaceous (and particularly the calcium carbonate in it) is an extremely good retarding material for the cationic species and delays migration of these elements by several decades.

f) Scenarios (fig. 14) such as a two fault scenario and a future unplugged borehole intersecting a disposal hole are considered [2]. The terrain at Mors is flat giving rise only to low potential heads. The driving force causing any flow is this low potential head from fresh water, and this is totally inadequate to sustain any large flows.

The density increase on the downstream side due to salt dissolution fully compensates this driving force. Even in large open fissures the flow will therefore be small and only sustained due to dispersion of the saline water into the matrix. Because of very low matrix permeabilities this flow will be minute. Calculations involving these scenarios have proved that although they represent unlikely and drastic conditions, they are quite harmless in their consequences [2].

g) Hypothetical questions regarding impossible geological and other events, which in no way are caused by the disposal of the radioactive waste in the dome, can always be asked. Such processes before they can reach the waste, however, will have to leach a substantial part of the dome surface. This will result in land subsidence. The Mors island will then disappear below sea level. Any potentials causing flow and leaching will thereafter decrease, and ultimately the flow will stop, keeping the radioactive waste still intact. The Mors salt dome with its island hydrology is dome preserving. That we have an island on top of a dome is proof enough.

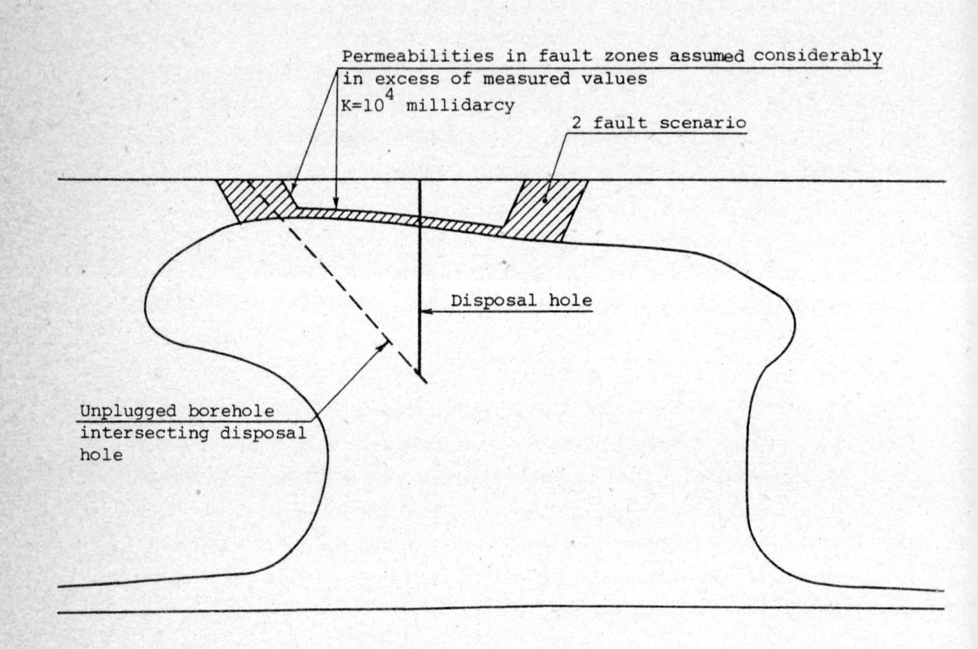

Typical scenarios

Figure 14

9. Conclusion.

Mors salt dome has around Erslev 1 and 2 large amounts of Zechstein 1 and 2 salt which is preminently well suited for radioactive waste disposal. A suitable hydrogeology is indicated by the tight caprock, low permeabilities, and low potentials in the formation above. The Cretaceous is very good at retaining radio-nuclides. The fresh and saline waters over the dome form an interface. The density pattern makes upwards flows above the caprock at Erslev 2 insignificant except by diffusion (and at the fresh-saltwater interface by dispersion and diffusion). The island location undoubtedly has advantages in the form of giving a simple hydrology which is dome preserving with its low potential gradients. There does not appear to be any reasonable mechanism which can bring the waste up, before several million years, when it no longer is radioactive.

10. Acknowledgements.

This paper is based on the work of many individuals, companies, and institutes - and my sincere thanks to all of them. The Geological Survey of Denmark has played a leading part in the geological investigations - and I have profited greatly through discussions with geologists from that institute. Søren Mehlsen from ELSAM has in times of difficulties - there have been many of them - given me backing, encouragement, and inspiration. For that I am very grateful. From my colleagues I have received much criticism - fortunately constructive and therefore always appreciated.

References.

1. Disposal of high-level waste from Nuclear Power Plants in Denmark, Vol 1 & 2, ELKRAFT and ELSAM, July 1978.

2. Disposal of high-level waste from Nuclear Power Plants in Denmark, Phase 2.
Vol 1 to 5, ELKRAFT and ELSAM (to be published in June 1981).

3. Erslev No 2. Well completion report. Finn Jacobsen and Steen Midtgaard Nielsen. Geological Survey of Denmark. Dec. 1980.

4. Geological remarks about the North Jutland Salt Domes in respect of their suitability for radioactive waste disposal. Prof. Dr. G. Richter-Bernburg 28.2.1981.

5. Hydrogeological Main Report.
E. Gosk, N. Bull, L.J. Andersen.
Geological Survey of Denmark April 1981.

6. Geological Main Report.
Geological Survey of Denmark April 1981.

7. Permeability, porosity, dispersion, and sorption characteristics of chalk samples from Erslev, Mors, Denmark.
Walther Batsborg, Lars Carlsen, and Bror Skytte Jensen.
Risø, Denmark. March 1981.

8. A hazard index for underground toxic material.
Craig F. Smith, Jerry J. Cohen, Thomas E. McKone, Lawrence Livermore Laboratory. VCRL-52889, June 1980.

9. Nuclear Wastes: A history of Deferrals.
Lawrence Livermore Laboratory. VCRL-82219, July 30, 1979.

10. Suggested Nuclear Waste Management Radiological Performance Objectives.
J. Cohen, Nureg/CR-0579, VCRL-52626, July 1979.

11. Risk Assessment and Radioactive Waste Management.
J.J. Cohen, VCRL-79293.

12. Statement to the Interagency Review Group on Nuclear Waste Management at San Francisco Ca.
J.J. Cohen, July 21, 1978.

13. Geochemische Prozesse in marinen Salzablagerungen: Bedeutung and Konsequenzen für die Endlagerung radioaktiver Substanzen in Salzdiapiren - Alberg Günter Herrmann.
Z.dt. Geol. Ges. 131. Hannover 1980.

14. Chemical analyses of pore water in chalk samples from Erslev, Mors, Denmark.
P. Solgaard and B. Skytte Jensen.
Risø - April 1981.

DISCUSSION

L.J. ANDERSEN, Denmark

Before we start the discussion on the prior papers, I have to comment the paper presented by Mr. Joshi, concerning the result of the Danish Investigation Programme.

This is due to the fact that the participants of this meeting could get the impression that the conclusions given in the paper - without reservations - are in agreement with the results obtained by the Geological Survey - and this is not the case.

Furthermore I feel that the countries represented here, seriously working with different aspects of geological disposal of radioactive waste during the last 2 to 3 decades, might be rather impressed by a Danish 1-2 years investigation programme able to solve all the waste disposal problems.

I am glad to recognise the presence of representatives of the Danish Environmental Agency, which body is in charge of approving the results of the investigations, described in this paper, and I have to inform that the results of this evaluation should be the base for the referendum concerning the introduction of nuclear power in Denmark.

B. HAIJTINK, CEC

You state in your paper that the salt is to be considered as the primary barrier. Nevertheless the design includes the encapsulation of the glass in a 15 cm thick steel cylinder. Do you consider this encapsulation as an additional artificial barrier and if so what is the expected lifetime ? If not, what is the reason of taking such a thick container ?

A.V. JOSHI, Denmark

Salt is the main geological barrier. The 15 cm steel overpack provides an engineering barrier. The reasons for this overpack are a) ease of handling and mechanical stability, b) reduction of radioactivity and temperature at salt interface, c) protection of glass against geostatic pressure, d) prevention of water reaching glass. The thickness is determined by (b) and (c). The question regarding expected lifetime is difficult to answer. The brine content in the Mors salt is not enough to give any brine migration and without oxygen from radiolysis of water there can be no corrosion. Hence I would expect a very long lifetime - several thousand years. However, we have worked with scenarios with very limited lifetime (2 vol. 5).

F. GERA, Italy

In relation to the curve showing the variation of rate of uplift of the Mors salt dome I would like to ask two questions :

1) How was the curve derived ?

2) How do you reconcile the apparent "rapid" (in a relative way) uplift of the last few thousand years with the very little salt left in the mother bed ?

A.V. JOSHI, Denmark

1) The curve showing the uplift of the surface of the salt dome at different times was based on the thickness of sedimentation layers and the uplifts observed. Information from seismic and regional geology was important.

2) The envelope to the dotted line indicates uplift velocities on a long term basis. Here the thickness of the salt left at the base is important. During the last ice age the edge of the Mors dome was covered by a very thick layer of ice. The uneven pressure perhaps gave higher uplift velocities. Thereafter the rebound from release of compression was also important. The total movement however was very small and the salt volume necessary to cause this uplift was therefore small.

J.D. MATHER, United Kingdom

You say that the homogeneous nature of the salt series found in Erslev 2 proves that the presence of significant amounts of K and Mg salts is unlikely. I wonder how the results from one borehole can be said to prove this ?

You give two possible reasons for the low yield of wells in formations overlying the salt dome and consider that the only plausible reason for this is that the old fissures are sealed by sedimentation and diagenesis. Could you say why you feel that the other explanation, put forward by the Geological Survey, that the central area of Mors is a compression zone, is not plausible ?

A.V. JOSHI, Denmark

Results of Erslev 1 and 2 show large amounts of Na1 and Na2 salt. Na1 does not contain significant amounts of K and Mg salts. These salts first appear between the boundary of Na2 and Na3. This is known from several drilled holes in the Danish bassin. The results of cores from Erslev 1 and 2 are complimented by electromagnetic and borehole gravimetric measurements. On this basis it is reasonable to expect that the necessary volume of salt free from significant amounts of K and Mg salts can be found in the Mors salt dome. Before actual disposal takes place, after year 2050, additional drilling will of course be necessary.

Of the two plausible reasons I consider the first reason - that low permeabilities in chalk at this depth are generally to be expected to be more likely. For discussion of question kindly refer to reference 2, vol. 2. What I believe would be helpful is to gather more information on the properties of chalk at the depths considered - particularly physical and geochemical. The cementing properties of chalk may be interesting to investigate.

L.J. ANDERSEN, Denmark

In addition to the remarks given by Dr. Mather (UK) I should like to say that the sentence on page 117 by Joshi : "I do not quike agree with that idea", seems to me a little nonchalant, as it is more than an idea. From the saltdomes in the North Sea-area there are several indications for the compression hypothesis, where compression zones are found in layers above areas of concave surface of the cap rock. Until this hypothesis is rejected the geologists would maintain it.

A.V. JOSHI, Denmark

I agree that the correctness of the hypothesis is difficult to prove or disprove at present. With all due respects to the geologists

I however beg to differ with their hypothesis - but of course I may be wrong. The surface of the Mors dome fig. 1 and 2 is not concave enough for a compression zone to exist and the wells S1 and S2, located as they are on the topmost part of the dome, fig. 3, would on the basis of the compression theory of geologists from the Geological Survey should be in the tension zone. The low permeability found in these wells however contradicts the possibility of a tension zone there. Neither below the bottom of the wells is there a possibility of a tension zone as the wells end almost above the caprock. You see the compression zone theory does not quite fit in with the picture of Mors as we know is today on the basis of seismic. Nor are there signs of leaching of the dome surface which the compression theory with a tension zone beneath will entail.

More information on permeabilities of chalk at these depths from other locations will perhaps show that the low permeabilities observed are the rule rather than an exception.

F. GERA, Italy

Could you give more details on the electromagnetic technique used to investigate the internal structure of the dome ?

A.V. JOSHI, Denmark

I would refer you to reference 2, vol. 2. The method has been developed at BGR by Dr. Tierbach and Nickel together with Prahla-Seismor. At Mors we could "see" like radar 600 m into the salt from the boreholes with the help of this method. Although the use of this method in boreholes is very new and has required the development of a new system, its use in mines is well known.

F. GERA, Italy

Have you considered the possibility of using microseismic techniques to lissen the possible differential movements inside the salt dome ?

A.V. JOSHI, Denmark

We have not considered that possibility. The movements in salt domes are very small and if measured in a borehole the cracking from the creep of the salt will perhaps influence the results.

UTILISATION DES DIAGRAPHIES* DANS LA DETERMINATION DES CARACTERISTIQUES LITHOLOGIQUES DANS L'ARGILE DE BOOM ET LEUR INTERET POUR LES CORRELATIONS DANS L'ARGILE[1]

B. Neerdael[2], N. Vandenberghe[3]
A. Bonne[2], E. Fierens[3], P. Laga[3], P. Manfroy[2]

RESUME

Le but du projet HADES en cours de réalisation au C.E.N./S.C.K. de Mol (Belgique) est d'abord rappelé ; il s'agit d'évaluer la possibilité de rejet des déchets radioactifs conditionnés dans la formation argileuse de Boom, située approximativement entre 160 et 270 m de profondeur.

Les étapes et aspects pris en considération lors de l'investigation du site sont passés en revue. Pour chacun d'entre eux, les différentes techniques mises en oeuvre sont mentionnées.

Les auteurs se penchent alors sur une de ces étapes, à savoir, l'extrapolation latérale des variations lithologiques, constatées lors de l'examen détaillé du sondage initial carotté. Parmi les méthodes applicables, la diagraphie de résistivité se révèle être, dans les sondages de reconnaissance, à la fois pratique et précise.

Enfin des exemples de corrélation très détaillée sont présentés aussi bien à l'échelle du site ou de l'unité hydrogéologique concernée, qu'à l'échelle régionale (existence de l'argile de Boom sur le territoire belge).

* bore logging

ABSTRACT

It is first recalled that the purpose of the HADES project of the C.E.N./S.C.K. at Mol (Belgium) is to study the possibility of geological disposal for radioactive wastes in the Boom clay formation at a depth between ± 160 and 270 m.

The different steps and aspects of the site investigation are exposed, and for each of them, the specific techniques used are reviewed.

One of these steps is elaborated more in detail. It covers namely the lateral extrapolation of the lithological variations as established from a very detailed analysis of the initial cored boring. Among the usefull methods, the most practical and refined results were obtained from the resistivity logs in the reconnaissance holes.

Finally, examples are shown of very detailed correlation on site scale, scale of the hydrogeological entity and regional scale (occurrence of Boom clay on the belgian territory).

[1] *Travail réalisé dans le cadre d'un contrat entre la Communauté européenne de l'Energie atomique et le Centre d'Etude de l'Energie nucléaire C.E.N./S.C.K.*
[2] *Centre d'Etude de l'Energie nucléaire C.E.N./S.C.K. MOL (Belgique)*
[3] *Service Géologique de Belgique SGB/BGD - BRUXELLES (Belgique)*

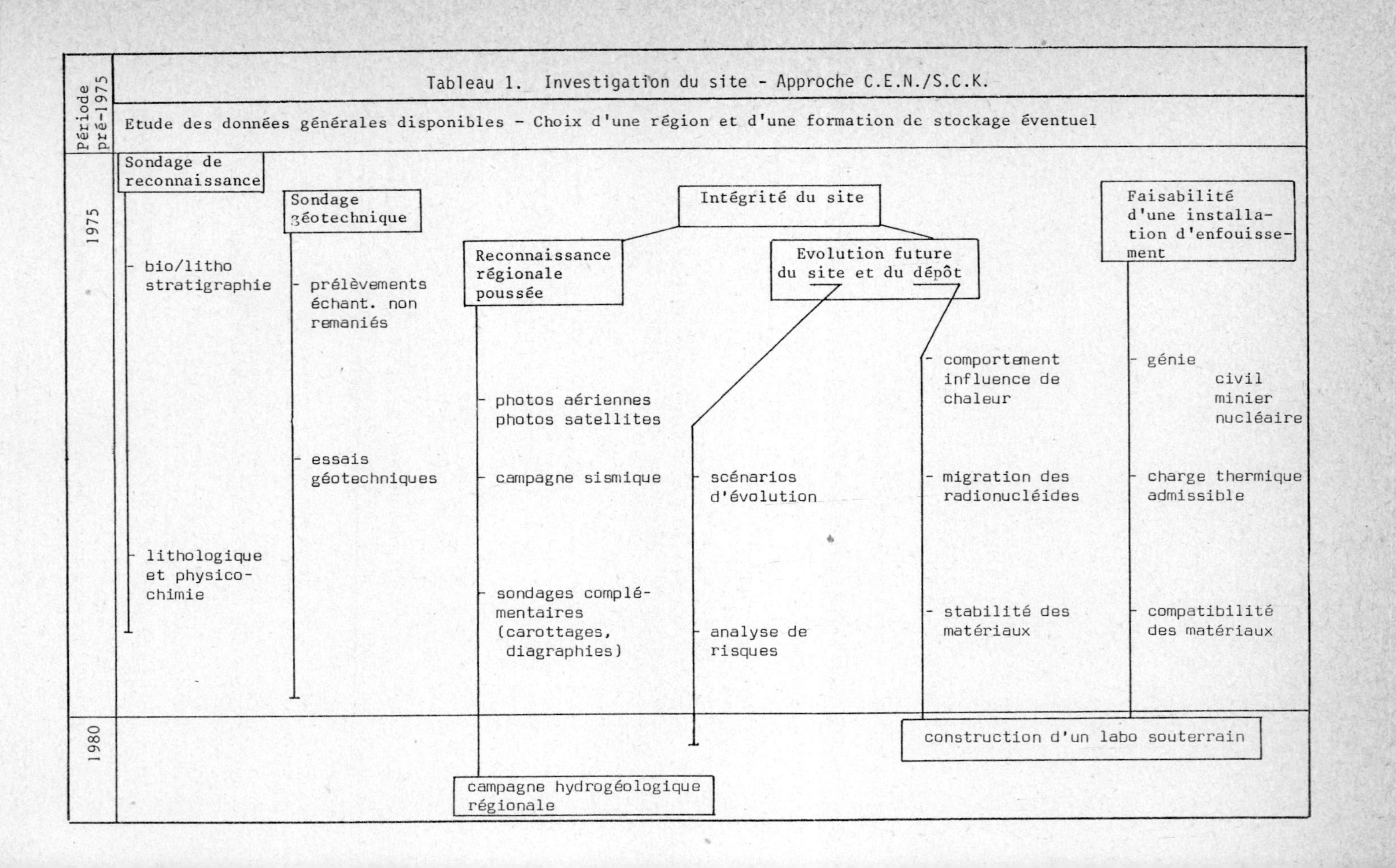

Tableau 1. Investigation du site - Approche C.E.N./S.C.K.
Période pré-1975
Etude des données générales disponibles - Choix d'une région et d'une formation de stockage éventuel
1975
Sondage de reconnaissance
bio/litho stratigraphie
lithologique et physico-chimie
Sondage géotechnique
prélèvements échant. non remaniés
essais géotechniques
Intégrité du site
Reconnaissance régionale poussée
photos aériennes photos satellites
campagne sismique
sondages complémentaires (carottages, diagraphies)
Evolution future du site et du dépôt
scénarios d'évolution
analyse de risques
comportement influence de chaleur
migration des radionucléides
stabilité des matériaux
Faisabilité d'une installation d'enfouissement
génie civil minier nucléaire
charge thermique admissible
compatibilité des matériaux
1980
construction d'un labo souterrain
campagne hydrogéologique régionale

1. INTRODUCTION

La mise en place progressive d'une industrie nucléaire en Belgique a amené le Centre d'Etude de l'Energie nucléaire (C.E.N./S.C.K.) de Mol à examiner les possibilités d'évacuation définitive de déchets émetteurs α et de moyenne ou haute radioactivité β, γ produits sur le territoire national.

Il est couramment admis dans les milieux scientifiques et techniques concernés, que des formations géologiques judicieusement sélectionnées peuvent jouer le rôle de barrière contre la dispersion dans la biosphère des radionucléides solidifiés qui y seraient entreposés.
Homogénéité, imperméabilité, plasticité et stabilité tectonique sont parmi d'autres des critères très importants à prendre en considération pour le choix d'une formation géologique destinée au rejet de déchets.

Un premier examen des formations profondes situées en Belgique et satisfaisant aux critères de sélection imposés montre que seules certaines couches schisteuses et certaines formations argileuses plastiques présentent des caractéristiques adéquates. L'étanchéité des premières n'est pas prouvée, même à grande profondeur, quant aux secondes, diverses couches relativement étendues et d'une épaisseur suffisante sont connues de longue date et situées à des profondeurs techniquement accessibles, entre autre dans la Campine Belge et en particulier, sous le site nucléaire de Mol.

Compte tenu de l'avantage que constitue la disponibilité d'une installation d'évacuation sous un site nucléaire existant, un programme détaillé de recherches a été élaboré par le C.E.N./S.C.K. pour aboutir à la mise en service d'une installation expérimentale souterraine au sein de la formation argileuse de Boom, présente, sous une forme compacte et homogène entre les niveaux approximatifs de 160 et 270 mètres.

Nous allons décrire brièvement l'approche du sujet telle qu'elle a été effectuée par le C.E.N./S.C.K. pour ensuite mettre l'accent sur l'intérêt que présentent dans le cas présent les diagraphies de résistivité pour l'interprétation des caractéristiques lithologiques dans l'argile de Boom.

2. APERCU DES DIFFERENTES ETAPES DE RECHERCHE ET TECHNIQUES D'INVESTIGATION UTILISEES

2.1. Aperçu général

Les différentes étapes de cette investigation ont été schématisées dans le tableau I, en relation avec le facteur temps.

On peut ainsi distinguer trois périodes majeures :

- *période antérieure à 1975* : conception du projet et étude des données disponibles pour le choix d'une région et d'une formation potentielle ;
- *période 1975-1979* :
  - étude de faisabilité d'une installation pour l'évacuation de déchets radioactifs conditionnés ;
  - reconnaissance locale et régionale détaillée par forages ;
  - essais de laboratoire variés sur échantillons carottés ;
  - étude de l'évolution future du site et du dépôt ;
- *période 1980-1982* :
  - construction du laboratoire souterrain pour la réalisation d'expérimentations in situ ;
  - développement d'une campagne hydrogéologique régionale.

2.2. Techniques d'investigation utilisées

Période antérieure à 1975

Dans cette phase préparatoire, ce sont essentiellement les techniques de cartographie qui ont permis d'analyser les caractéristiques, notamment géométriques, des formations potentielles d'évacuation. La figure 1 montre, à titre d'exemple, l'une des coupes géologiques de ce type.

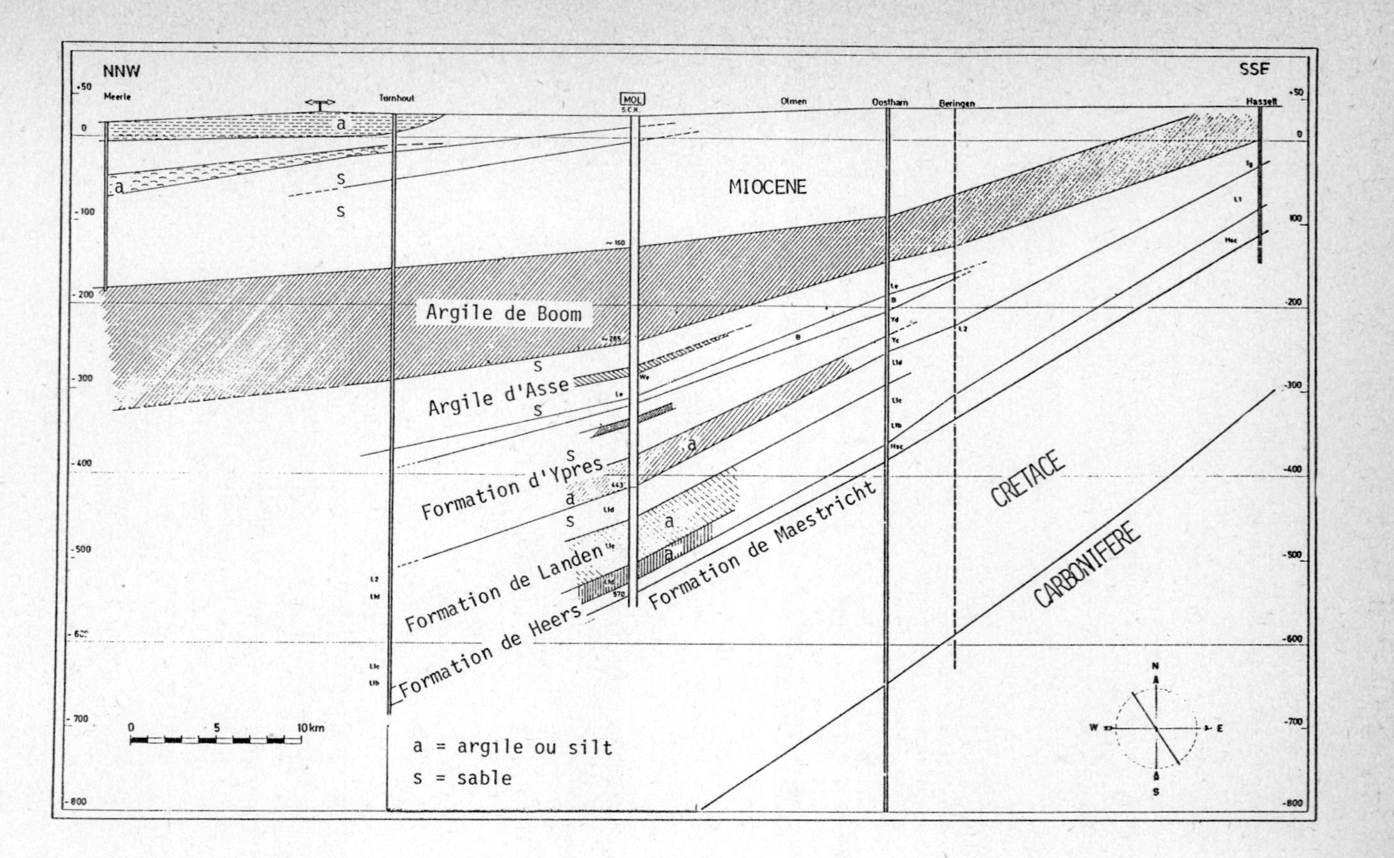

FIG. 1 : PROFIL GEOLOGIQUE NNW-SSE (MEERLE - HASSELT)

Période postérieure à 1975

2.2.1. Sondage de reconnaissance

Un forage carotté, complété par diagraphies et échantillonnage des différentes formations, a permis de dresser une description litho-stratigraphique détaillée des formations traversées (Fig. 2). La technique employée consiste en un carottier à câble enfoncé hydrauliquement dans le sol, autorisant le prélèvement continu des échantillons sur plusieurs centaines de mètres.

Les échantillons ainsi prélevés ont conduit à diverses déterminations de laboratoire sur l'argile de Boom dont les principales sont énumérées ci-après :

- analyses chimiques quantitatives (éléments majeurs et en trace, eau extraite, capacité d'échange, ...) ;
- caractéristiques géotechniques (densité, granulométrie, limites d'Atterberg,.) ;
- analyses minéralogiques de la fraction fine ;
- caractéristiques thermiques (conductibilité, chaleur spécifique, dilatation thermique, thermogravimétrie, ...).

Enfin, les diagraphies ont fourni des renseignements sur la résistivité, la radioactivité naturelle et la température des formations traversées, et, indirectement, sur la perméabilité des couches poreuses et la salinité des eaux.

2.2.2. Sondage géotechnique

D'autres sondages ont été effectués à proximité ; ceux-ci ont atteint les nappes aquifères les plus importantes afin d'y installer des piézomètres. Pour l'un de ces forages, un dispositif spécial a été développé à la demande des géotechniciens pour prélever des échantillons de sol présentant un remaniement minimum.

Ce système assurait une contre-pression au-dessus de la carotte lors de son prélèvement mais nécessitait par contre la remontée du train de tiges pour la récupération de l'échantillon. Sur ces échantillons d'un diamètre de 6", ont été réalisées les mesures du poids volumique, de la teneur en eau naturelle, des caractéristiques de perméabilité et de déformabilité et de la résistance au cisaillement.

2.2.3. Reconnaissance régionale

Afin de définir plus précisément les caractéristiques géométriques de l'argile de Boom ainsi que l'homogénéité latérale de la formation, diverses recherches et études ont été menées : il s'agit essentiellement d'une étude de photos aériennes et satellites ; d'une campagne sismique réflexion, et de sondages complémentaires.

Interprétation de photos satellites et photos aériennes

Une étude de photos satellites (LANDSAT) (images positives, négatives, superpositions, filtrages, couleurs composites, ...) faite en collaboration avec le Centre Commun de Recherches d'Ispra (CCE), ainsi que l'examen détaillé de la couverture photographique aérienne régionale, a mis en évidence la présence de linéaments dans le NE de la Belgique dont certains peuvent être la manifestation en surface de structures de discontinuités situées en profondeur.
Les directions ainsi identifiées ont été comparées aux données techniques déjà disponibles [1].

Campagne sismique réflexion de haute résolution

Cette campagne dite "en nappe" ou tridimensionnelle, a été entreprise pour le compte du C.E.N./S.C.K. par la firme PRAKLA SEISMOS d'Hanovre afin de déterminer de façon précise les caractéristiques géométriques de la couche d'argile de Boom, ainsi que leurs variations et de détecter toute structure particulière (tectonique, sédimentaire,...) susceptible de se présenter tant au sommet du socle paléozoïque que dans les terrains méso- et cénozoïque de couverture.

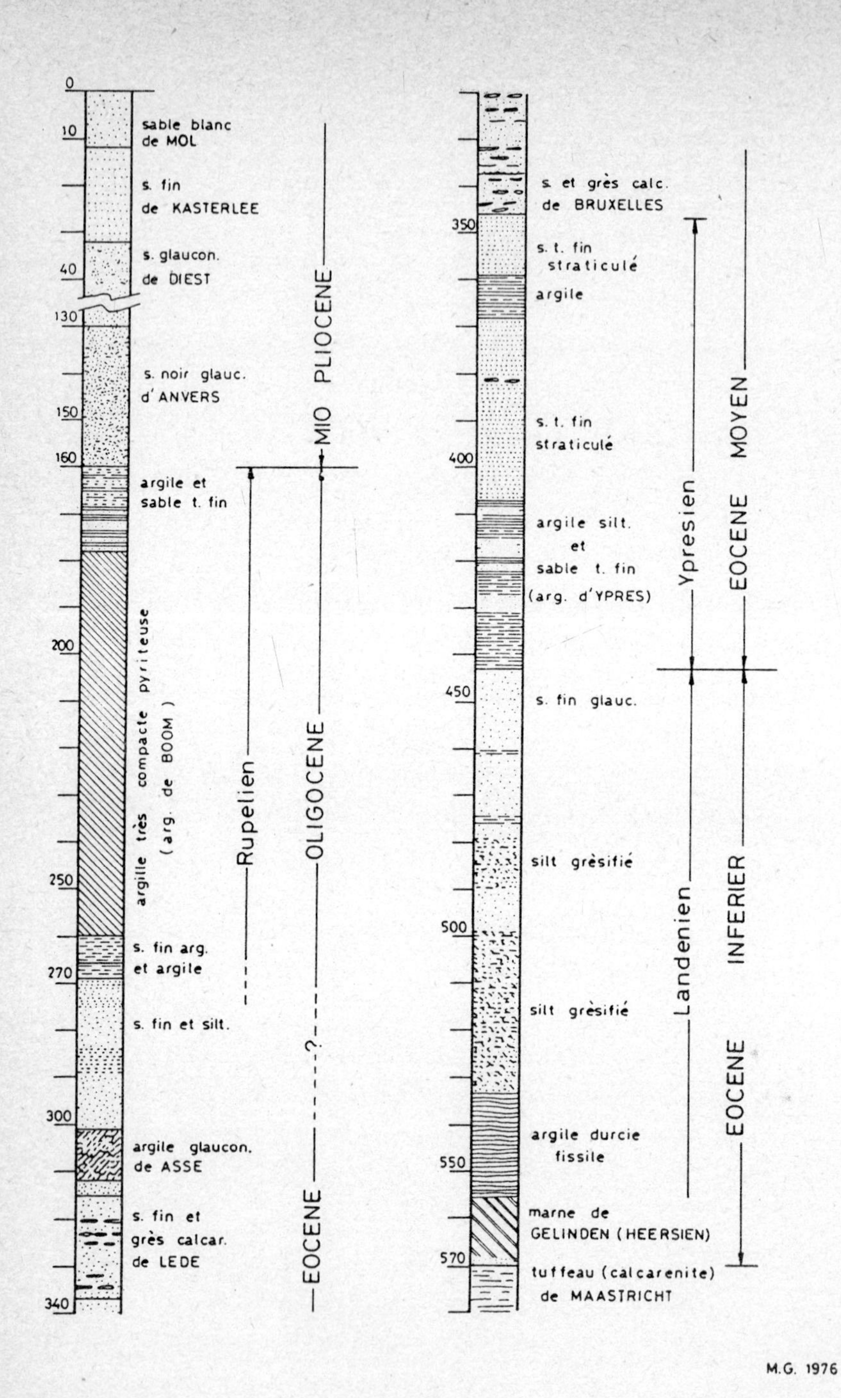

Fig.: 2 Coupe stratigraphique simplifiée du forage geologique.

Les paramètres d'acquisition utilisés étaient :

- temps d'enregistrement de 2 secondes ;
- couverture d'ordre 5 ;
- pas d'échantillonnage de 1 milliseconde ;
- système d'enregistrement télémétrique à 60 canaux et transcodage analogique-digital ;
- géophones à filtre passe-bas de 20 Hz en grappe de 18 ;
- source dynamite de 100 g à 2 m de profondeur.

Lors du traitement, les fréquences retenues par filtrage étaient de 30 à 250 Hz. Ces paramètres ont permis d'atteindre une résolution verticale du toit et du mur de l'argile de 2 à 3 mètres et d'obtenir des renseignements significatifs sur les terrains sous-jacents. A noter qu'un sismo-sondage préalable avait permis de préciser les paramètres de conversion des temps en profondeurs. La campagne a finalement confirmé la continuité tant en extension qu'en épaisseur de la formation argileuse de Boom , son léger pendage ($\simeq$ 1 %) en direction N-NE ainsi que l'absence de failles ou de flexures.

Sondages complémentaires

Afin de confirmer l'extension latérale des formations et leurs caractéristiques, les sondages réalisés dans la région pour le C.E.N./S.C.K. ont été suivis avec soin.

Les sondages congélateurs forés sur le site expérimental pour le creusement par congélation du puits d'accès à la galerie expérimentale, recoupent l'argile de Boom sur plus de la moitié de son épaisseur. Carottages et diagraphies ont permis de multiplier et de préciser les corrélations lithostratigraphiques.

En collaboration avec le Service Géologique, le C.E.N./S.C.K. a développé une campagne hydrogéologique régionale visant à caractériser en détail les nappes phréatiques et semi-phréatiques sus-jacentes à l'argile de Boom (niveaux piézométriques, compositions chimique et radiochimique influence des précipitations,..) elle contribue dans une large part à la reconnaissance détaillée de la région. 3500 m de forage permettront l'implantation d'une cinquantaine de piézomètres ; ceux-ci sont situés le long de deux profils intersectant la crête de partage de deux bassins fluviaux (Fig. 3).

Les résultats de cette campagne fourniront les renseignements nécessaires à l'établissement d'une modélisation hydrologique détaillée de toute la région.

Nous reviendrons au paragraphe suivant sur les corrélations qui ont été rendues possibles sur base notamment des diagraphies de résistivité.

2.2.4. Evolution future du site et du dépôt

Les travaux relatifs à l'étude de la stabilité géologique future doivent aboutir à l'évaluation de la sécurité à long terme.

Connaissant la nature et les vitesses des processus géologiques à prendre en considération, des scénarios d'évolution géologique ont été élaborés sur base des processus antérieurs et de leurs caractéristiques [2].

Une analyse probabiliste,par la méthode de l'"arbre de défaillance" [3] a été réalisée.Sur cette base, plusieurs scénarios probabilistes de libération, sous l'influence de sollicitations externes, ont été établis [4].

D'autre part, en ce qui concerne le dépôt lui-même, on se préoccupe encore actuellement de l'étude des scénarios d'évolution du dépôt, suite aux interactions chimiques, thermiques et mécaniques entre déchets, matériaux de structure ou d'emballage, ... et la formation elle-même.

2.2.5. Etude de faisabilité d'un complexe pour le stockage de déchets radioactifs conditionnés

Les essais de laboratoire énumérés en 2.2.1. et 2.2.2. représentent les données de base essentielles à l'étude de faisabilité du projet. Ainsi, les

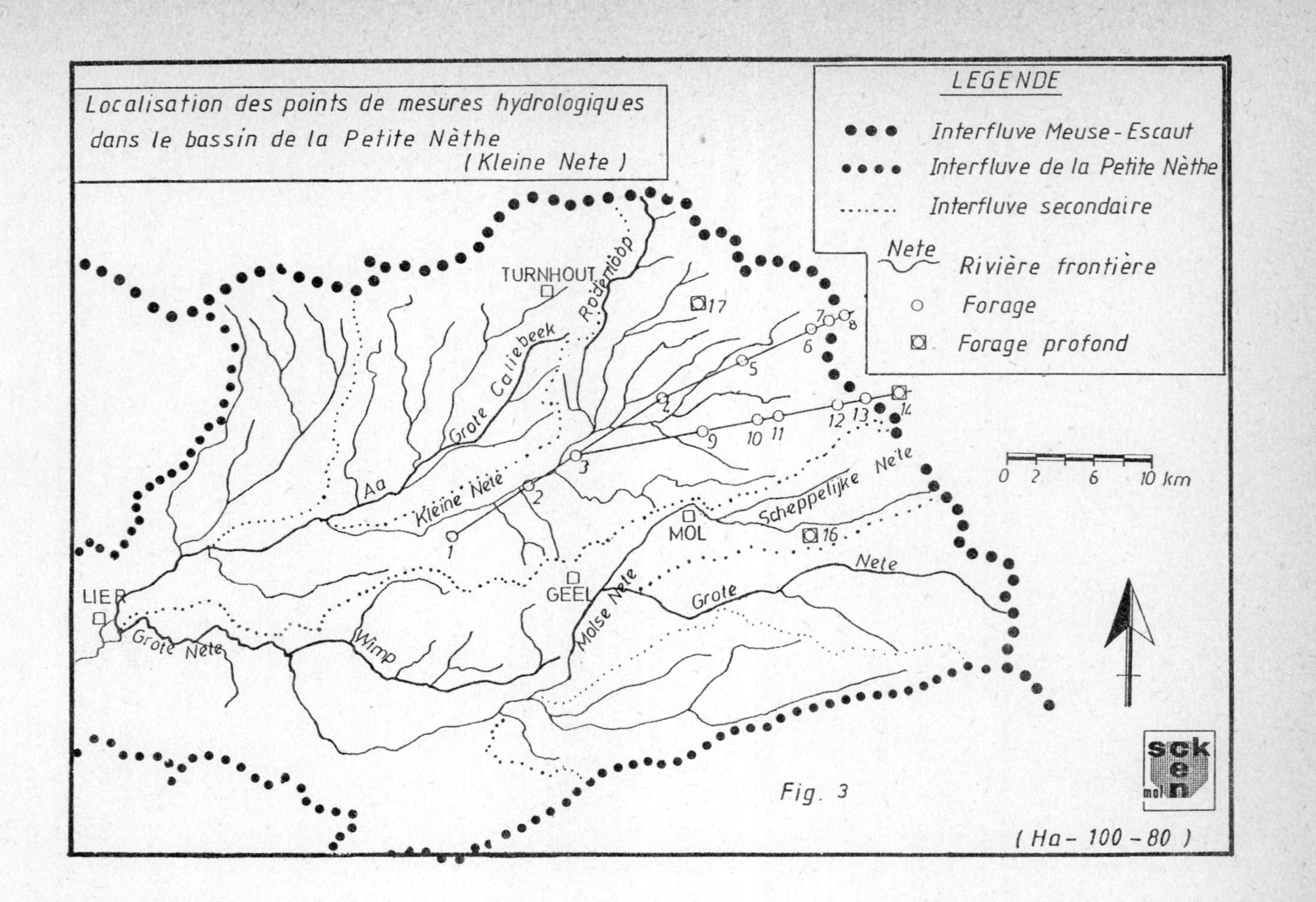
Localisation des points de mesures hydrologiques
dans le bassin de la Petite Nèthe
(Kleine Nete)
LEGENDE
Interfluve Meuse-Escaut
Interfluve de la Petite Nèthe
Interfluve secondaire
Nete Rivière frontière
Forage
Forage profond
0 2 6 10 km
TURNHOUT
Grote Caliebeek
Aa
Kleine Nete
Scheppelijke Nete
MOL
GEEL
Molse Nete
Grote Nete
Wimp
LIER
1 2 3 4 5 6 7 8 9 10 11 12 13 14 16 17
sck cen mol
(Ha-100-80)

Fig. 3

essais géomécaniques étaient indispensables pour concevoir la technique de réalisation des galeries souterraines. Quant à la détermination de la charge thermique admissible ; elle conditionne les dimensions des installations souterraines d'enfouissement à la profondeur envisagée; un modèle thermique est d'ailleurs à l'étude.

Les résultats obtenus ont confirmé la faisabilité du projet et ont rendu évidente la nécessité de construire un laboratoire expérimental souterrain afin de vérifier les hypothèses et évaluer les inconnues principales.

Cette galerie de 25 m de long, au sein de l'argile, permettra de réaliser, dans les conditions réalistes, les essais chimiques, métallurgiques, géotechniques et thermiques, jusqu'ici exécutés au laboratoire ou limités à des expérimentations in situ à faible profondeur (carrières de la région anversoise où l'argile affleure).

Les valeurs expérimentales ainsi relevées permettront, par comparaison avec les valeurs théoriques calculées, la mise au point des modèles mathématiques précédemment élaborés.

3. LES VARIATIONS LITHOLOGIQUES DANS L'ARGILE - LEUR DETECTION PAR DIAGRAPHIE DE RESISTIVITE

3.1. Existence de variations lithologiques au sein de l'argile compacte

Dans la région d'Anvers où l'argile de Boom affleure, des variations lithologiques d'une épaisseur de quelques dizaines de cm, exprimant le plus souvent des variations de teneur en silt, argile, carbonates et matière organique végétale, ont été mises en évidence et correlées d'argilière en argilière dans toute la région (Fig. 4). [5]

Cette microstratigraphie a aussi été retrouvée dans les sondages carottés par un examen minutieux des échantillons.
C'est notamment le cas d'un chantier situé au Nord-Ouest d'Anvers qui a conduit à l'établissement de la figure 5. [6]

On a procédé de la même manière pour compléter le log lithographique (établi en 1976) par le repérage précis de toutes les bandes d'argile plus silteuses, si petites soient-elles, (réexamen des carottes du sondage initial).
Il est apparu qu'une correspondance semblait exister entre les niveaux ainsi répertoriés et les très faibles variations de résistivité enregistrées sur le log normal au sein de l'argile.

3.2. Confirmation d'une corrélation avec le log de résistivité

Nanti de ces premières informations, les investigations ont été poursuivies par un examen similaire du second sondage carotté, réalisé en 1980 à l'emplacement d'un tube congélateur sur le site expérimental. Il a lui aussi confirmé l'existence de ces variations lithologiques régulières et leur correspondance avec la diagraphie de résistivité.

Les considérations et figures qui s'y rapportent appelent la remarque suivante :

- le forage initial carotté, de coordonnées 51°12'45" N et 5°3'37" E, avait fait l'objet de diagraphies de divers paramètres réalisés par la firme britanique WRC ; la sonde de résistivité utilisée, dite normale, possédait des espacements entre électrodes de 40 et 160 cm ;
- le sondage "congélateur"carotté, ceinturant le puits d'accès du projet expérimental au point de coordonnées 51°12'50" N et 5°5'15" E, situé à ± 2 km du précédent, a fait l'objet de diagraphies du même type réalisées par la firme néerlandaise T.N.O., les sondes utilisées pour la résistivité étaient à disposition dite normale et à espacement d'électrodes de 20, 40, 100 et 200 cm.

Ces précisions s'avèrent indispensables pour justifier le décalage d'une vingtaine de mètres existant au sommet de l'argile sur ces deux diagraphies et pour souligner l'intérêt de pouvoir corréler les informations provenant de deux firmes distinctes utilisant un matériel sensiblement différent.

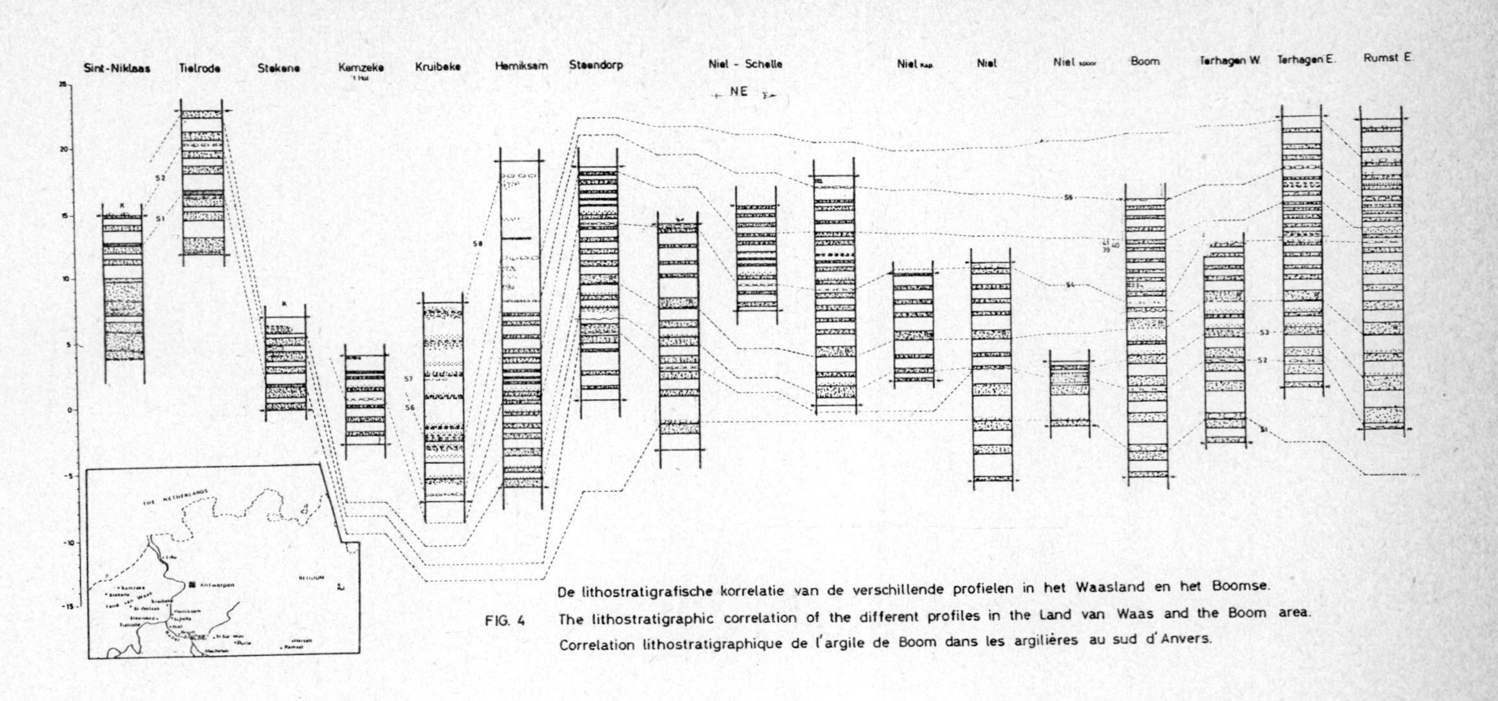

FIG. 4 De lithostratigrafische korrelatie van de verschillende profielen in het Waasland en het Boomse.
The lithostratigraphic correlation of the different profiles in the Land van Waas and the Boom area.
Correlation lithostratigraphique de l'argile de Boom dans les argilières au sud d'Anvers.

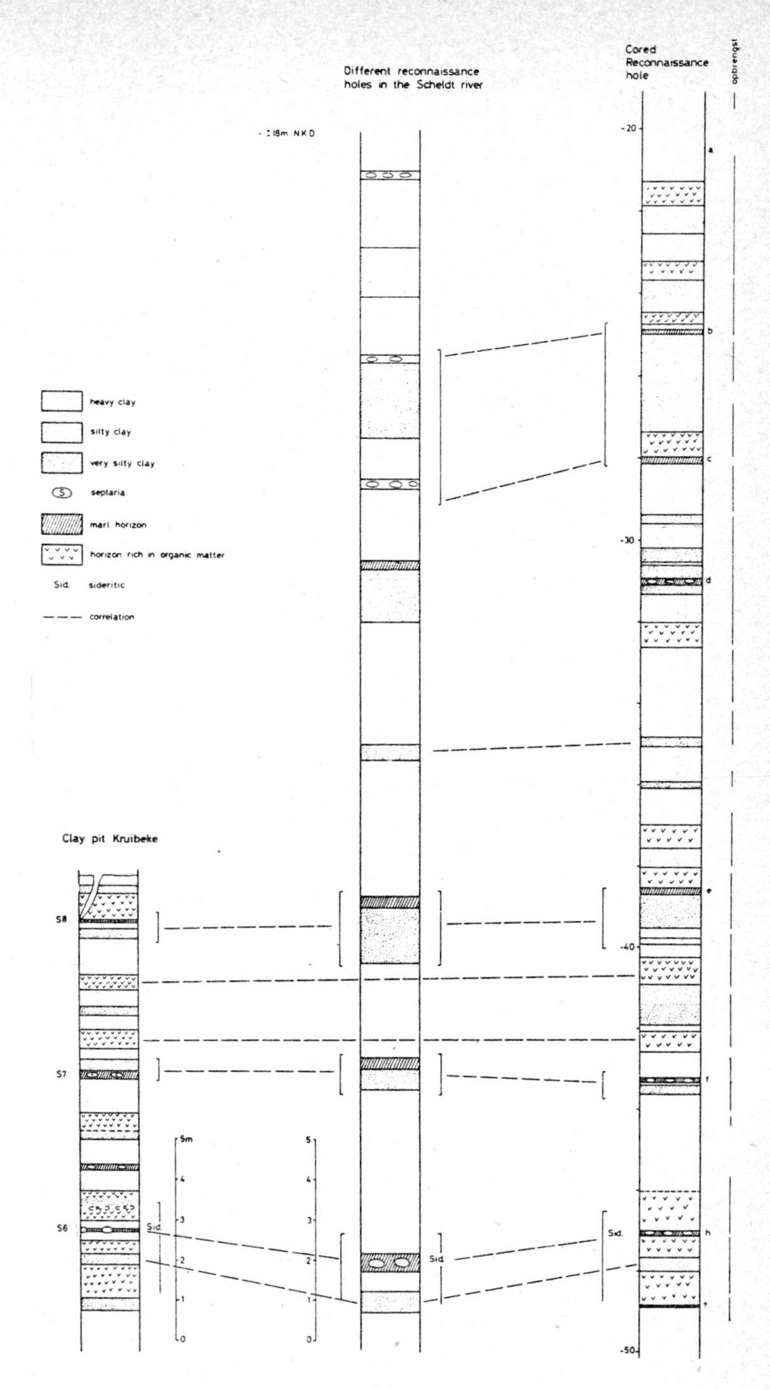

Fig. 5 Microstratigraphic correlation in the Boom clay between the cored and other reconnaissance holes in the Scheldt river and the clay pit at Kruibeke

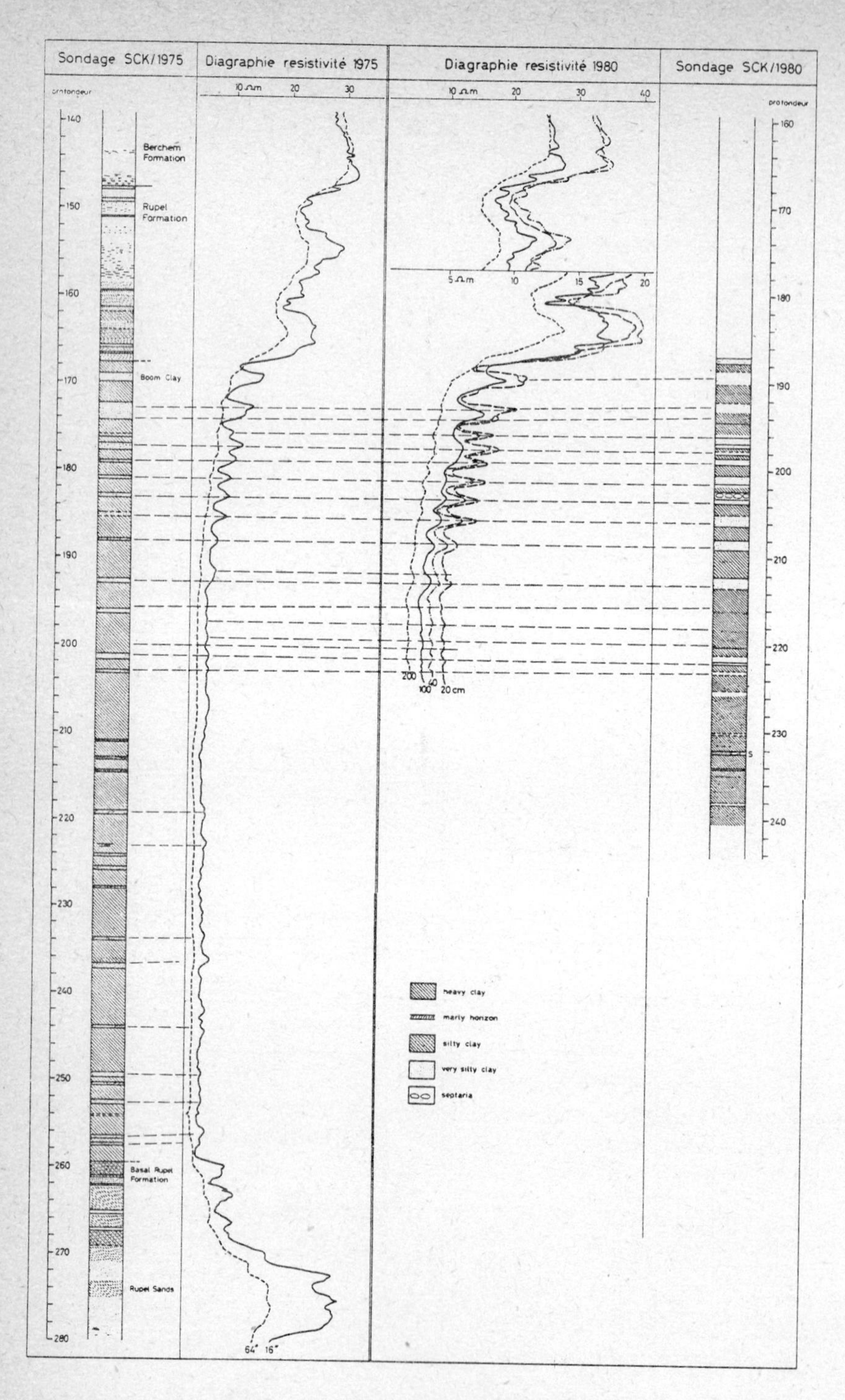

Fig. 6 Comparaison entre les deux sondages carottés au CEN à Mol et les deux diagraphies de résistivité correspondantes

La figure 6 représente les descriptions lithologiques et les courbes de résistivité des deux sondages en question ; la zone de transition du sommet de l'argile de Boom, caractérisée par neuf pics de résistivité répartis à Mol sur une vingtaine de mètres, se compose de passages relativement silteux qu'un examen succinct des carottes met nécessairement en évidence.
Quant à l'argile compacte, les commentaires suivants peuvent être faits :

- les sondes à électrodes espacées de 100, 160 et 200 cm intègrent une épaisseur d'argile trop grande pour différencier les minces bancs d'argile plus silteuse ;

- par contre, les espacements de 20 et même 40 cm permettent cette différenciation ; la correspondance entre les différentes interprétations est remarquable ;

- dans la partie supérieure, les variations de l'ordre de 0,5 Ωm indiquent des phénomènes répétables, correspondant à des variations très fines de la teneur en silt ;

- dans la partie inférieure (220-260 m), la courbe de résistivité du sondage initial montre une correspondance moins systématique entre variations de lithologie et de résistivité mais non vérifiable à cette profondeur sur le second forage. L'imprécision sur la profondeur carottée et la faible épaisseur des bancs plus silteux vis-à-vis de l'espacement des électrodes expliquent partiellement ce manque de correspondance que l'utilisation d'une sonde à espacement d'électrode de 20 cm aurait pu améliorer.

3.3. Corrélations particulières à la zone de transition

On s'est attaché dans cette zone à quantifier par la géophysique et la géotechnique les renseignements qualitatifs que l'observation visuelle des carottes pouvait donner quant au caractère plus ou moins silteux de l'argile.

Des échantillons prélevés sur les carottes du sondage "congélateur" ont été soumis aux essais classiques de granulométrie-sédimentométrie et limites d'Atterberg. Les résultats ainsi obtenus montrent une correspondance entre le caractère plus ou moins silteux du matériau prélevé, l'enregistrement correspondant de la résistivité et les caractéristiques géotechniques de l'échantillon en question mesurées en laboratoire.

Cette corrélation est telle que le diagramme de résistivité pourrait être gradué en abscisse pour la plupart des paramètres géotechniques mesurés.
A titre d'exemple la figure 7 illustre cette particularité pour le passant à 2 µ représentant donc la teneur en argile. Pour rappel, la résistivité est égale ou inférieure à 10 Ωm pour les argiles alors que cette valeur est de l'ordre de 30 Ωm pour un sable.

3.4. Corrélations à l'échelle de l'unité hydrologique

Les considérations développées dans les paragraphes précédents conduisent à la conclusion que les mesures géophysiques et notamment, les diagraphies de résistivité, effectuées dans une configuration géologique similaire, permettent la détermination des variations lithologiques au sein d'une formation argileuse ou encore une interprétation correcte de la nature et des caractéristiques des formations sableuses et argileuses rencontrées.

Il pourrait en être de même de logs γ naturel, tout aussi représentatifs de ce type de terrain mais, dans notre cas, la présence de glauconie en quantité importante dans les sables rend son interprétation beaucoup plus difficile.

Nous avons déjà eu l'occasion de nous pencher sur les corrélations possibles à l'échelle locale et régionale.
Dans la même optique, la figure 8, par la comparaison des diagraphies de résistivité à Retie et Balen, illustre particulièrement bien la régularité de la zone de transition au sommet de l'argile de Boom pour deux des forages hydrogéologiques ayant atteint la formation et situés à des distances de 10 à 15 km.
La figure peut pratiquement se passer de commentaires; nous ferons simplement remarquer que ces deux zones, tout en étant d'épaisseurs sensiblement différentes, présentent le même nombre de pics et, pour chacun de ceux-ci, une configuration rigoureusement identique.

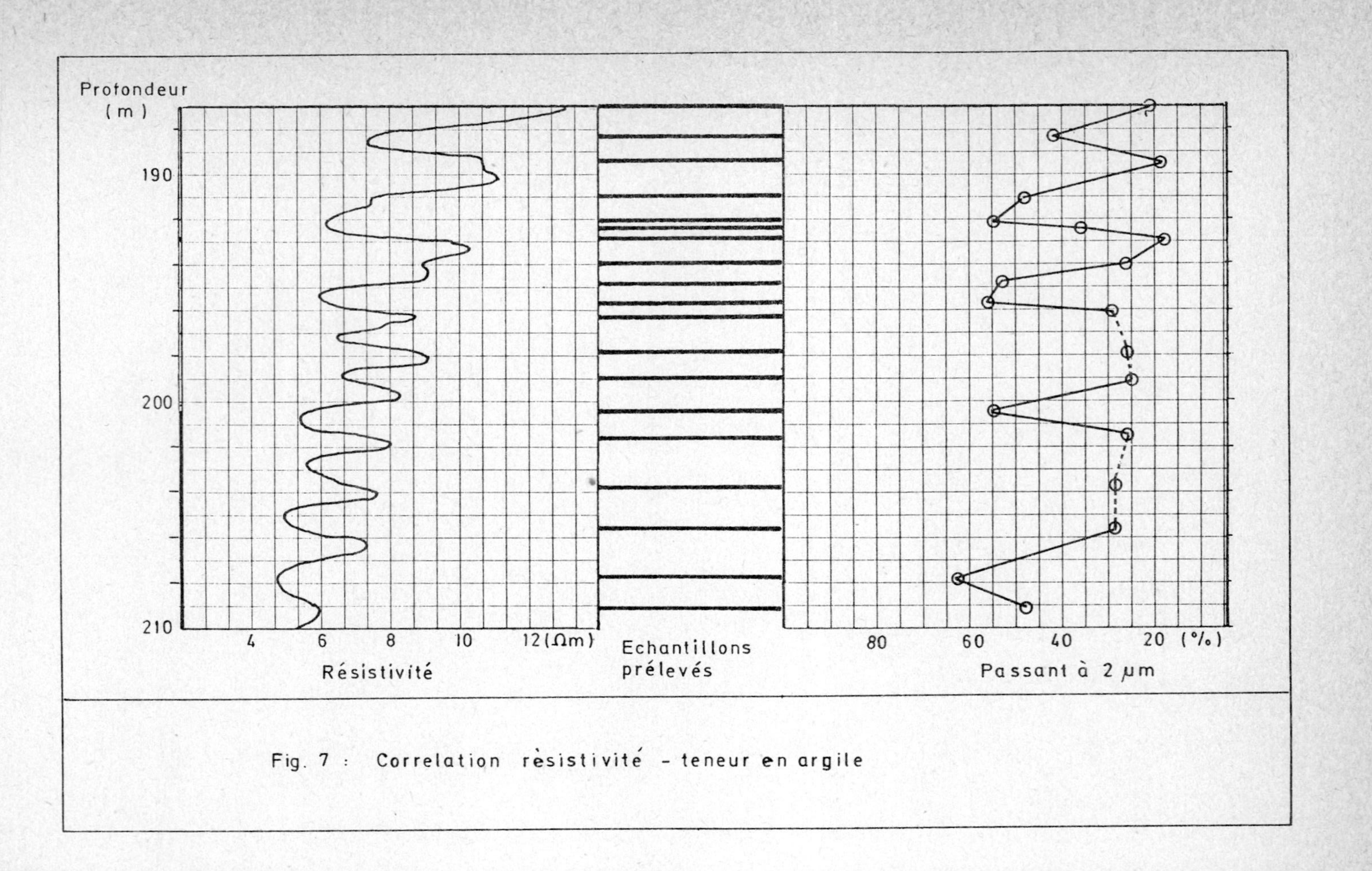

Fig. 7 : Correlation rèsistivité - teneur en argile

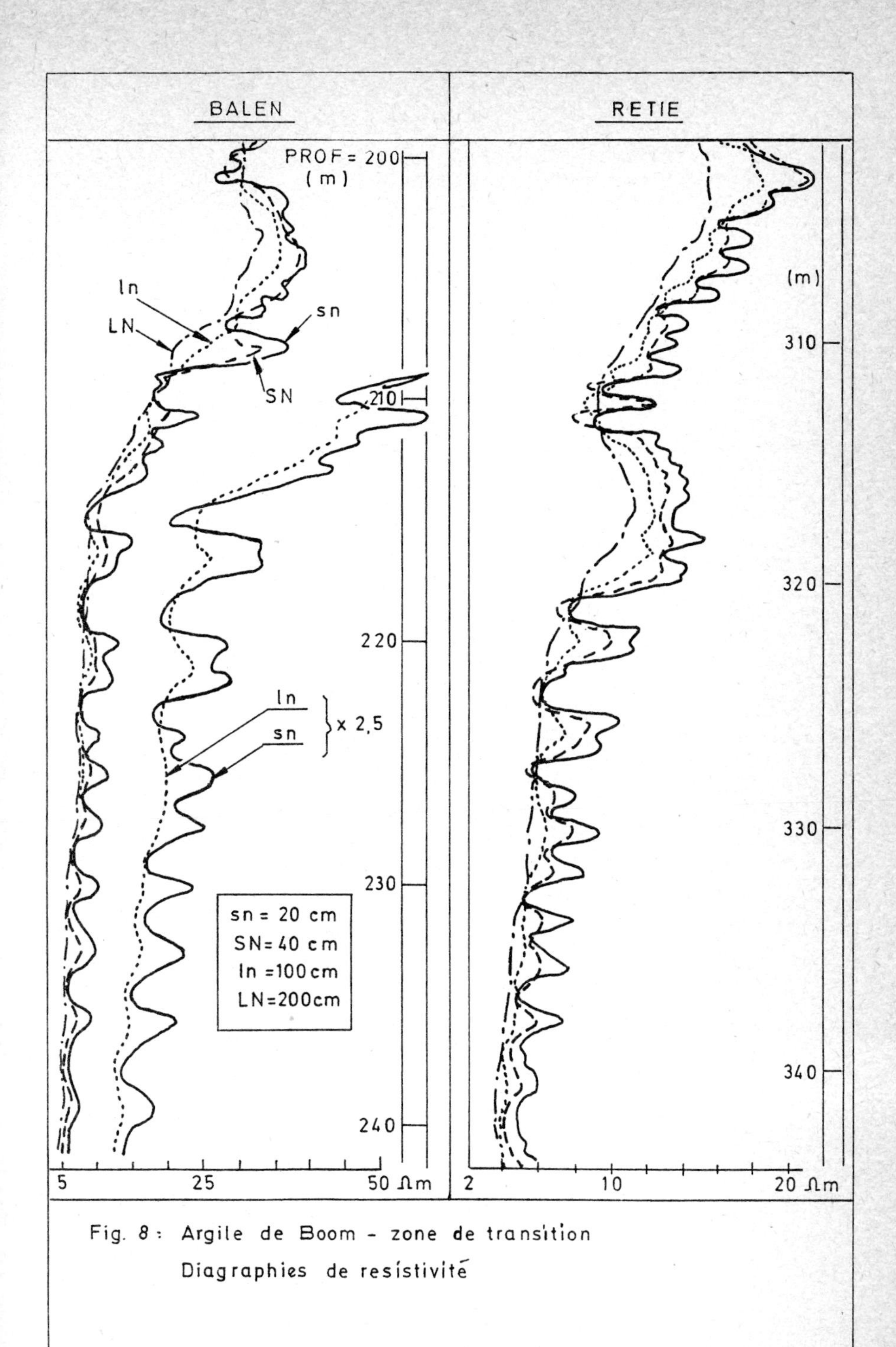

Fig. 8 : Argile de Boom - zone de transition
Diagraphies de resístivité

4. CONCLUSIONS

En conclusion, quelques particularités pratiques de la méthode appliquée doivent être signalées, à savoir :

- la quantité impressionnante de renseignements pouvant être déduits de cette diagraphie, réalisée dans le forage terminé (train de tiges remonté) mais non tubé et rempli de boue ;
- la précision des déterminations ainsi rendues possibles ;
- la simplicité du matériel mis en oeuvre, ce qui en fait un moyen de prospection peu onéreux.

Enfin, l'expérience décrite ici brièvement doit pouvoir être généralisée à toute formation argileuse ; l'existence de ces niveaux silteux plus ou moins abondants, stratiformes ou lenticulaires, est fréquente dans ces formations dites compactes et homogènes. Les diagraphies électriques classiques, dont le dispositif normal à faible espacement d'électrodes, se révèle être un moyen de prospection particulièrement adapté au problème ainsi qu'une technique de corrélation très puissante.

REFERENCES

1. Bonne, A., Heremans, R., Manfroy, P., Dejonghe, P. : "Investigations entreprises pour préciser les caractéristiques du site argileux de Mol comme lieu de rejet souterrain pour les déchets radioactifs solidifiés", in : Underground Disposal of Radioactive Wastes, Vol. 2, 41-58, Int. Atom. Energy Agency, Vienna, 1980, (ISBN-92-0-020280-2).

2. Vandenberghe, N., Bonne, A., Heremans, R. : "Scénarios d'évolution géologique lente appliqués au site argileux de Mol (Belgique), in : Proc. Workshop on Radionuclide Release Scenarios for Geologic Repositories, Paris, Septembre 8-12, 169-180, Nucl. Energ. Agency, OCDE, Paris, 1980 (ISBN-92-64-02172-8).

3. D'Alessandro, M., Bonne, A. :"Fault Tree Analysis for Probabilistic Assessment of Radioactive Waste Segregation. An Application to a Plastic Clay Formation at a Specific Site", in : Predictive Geology, Proc. Int. Geological Congress, Paris, July 1-17, Pergamon Press, 1980.

4. D'Alessandro, M., Bonne, A. : "Radioactive Waste Disposal into a Plastic Clay Formation : Probabilistic Assessment of the Geological Containment" in : Scientific Basis for Nuclear Waste Management, Vol. 2, 711-720, Plenum Press, New-York and London, 1980 (ISBN-0-306-40550-4).

5. Vandenberghe, N. : "Sedimentology of the Boom Clay (Rupelian) in Belgium", Compte Rendus de l'académie royale des sciences, lettres et beaux-arts de Belgique, n° 147, 1978.

6. Fierens, E. :"Bijdrage tot de kennis van de Geologie van het site voor de stormstuw in de Schelde te Oosterweel (Antwerpen), Adm. Mijnwezen, Min. Econ. Zaken, Geol. Dienst, Rapport.

DISCUSSION

A. BRONDI, Italy

Vous avez démontré que les mesures de résistivité ont mis en évidence une grande régularité et ont confirmé l'extension latérale sur une échelle régionale des intercalations silteuses dans les argiles pupeliennes en Belgique. Est-ce que cette régularité est due à des conditions de formation identiques dans la région ? Quel était le degré d'évolution morphologique du continent au moment de ce dépôt de sédiments effectué de façon si régulière ?

N. VANDENBERGHE, Belgium

La régularité latérale est certainement due aux conditions de génèse. On a pu démontrer dans la zone d'affleurement de l'argile de Boom que l'alternance de couches très argileuses et plus silteuses est probablement causée par la possibilité des vagues de remuer périodiquement les sédiments au fond de la mer. Comme la profondeur d'activité des vagues dépend de la longueur des ondes, des variations climatiques peuvent l'avoir influencée ; néanmoins je pense que la mobilité tectonique du fond de mer ou des effets de tectono-eustatisme ont changé périodiquement la profondeur de la mer épicontinentale, une profondeur estimée à environ 50 m.

Il faut encore ajouter que les couches silteuses identifiées dans la région de Mol n'ont pas encore pu être corrélées avec la microstratigraphie au sein de l'argile dans la zone d'affleurement. Il est déjà sûr que toutes les couches silteuses ne pourront pas être tracées jusqu'à la région de Mol.

Il est difficile de juger de la morphologie du continent tout au moins d'une façon précise. Pendant l'oligocène il y a eu d'importants mouvements tectoniques alpins qui ont provoqué la surrection de la partie méridionale du bassin sédimentaire de sorte que la ligne côtière miocène se trouve déjà beaucoup plus au nord.

A. BRONDI, Italy

Existe-t-il des variations minéralogiques importantes dans toute l'épaisseur d'argile et dans un même niveau ?

N. VANDENBERGHE, Belgium

Sauf quelques variations mineures, la réponse est négative tout au moins dans la région d'affleurement. Les différences minéralogiques de la fraction argileuse observées dans le sondage de Mol en 1975 sont, à mon opinion, causées par la technique d'analyse différente.

IDENTIFICATION OF SUITABLE GEOHYDROLOGIC ENVIRONMENTS FOR THE DISPOSAL OF HIGH-LEVEL RADIOACTIVE WASTE

M. S. Bedinger and K. A. Sargent
U.S. Geological Survey
Lakewood, Colorado

ABSTRACT

The U.S. Geological Survey's effort to identify favorable locations for mined repositories will initially involve 1 of 11 provinces encompassing the conterminous United States.

The screening process will be accomplished by the U.S. Geological Survey in conjunction with a Province Working Group comprised of earth scientists from the Geological Survey and the States in the Province.

Geologic and hydrologic evaluation of the Province will be based on an adopted set of screening guidelines that stress factors affecting the isolation of waste from the accessible biosphere through a system of multiple natural barriers. The first stage of screening will emphasize definition of ground-water flow systems and distribution and characteristics of potential host rocks, including evaporties, tuffs, basalts, crystalline igneous rocks and shales. The essential attribute of the flow system is to provide long residence time before the water enters the accessible environment.

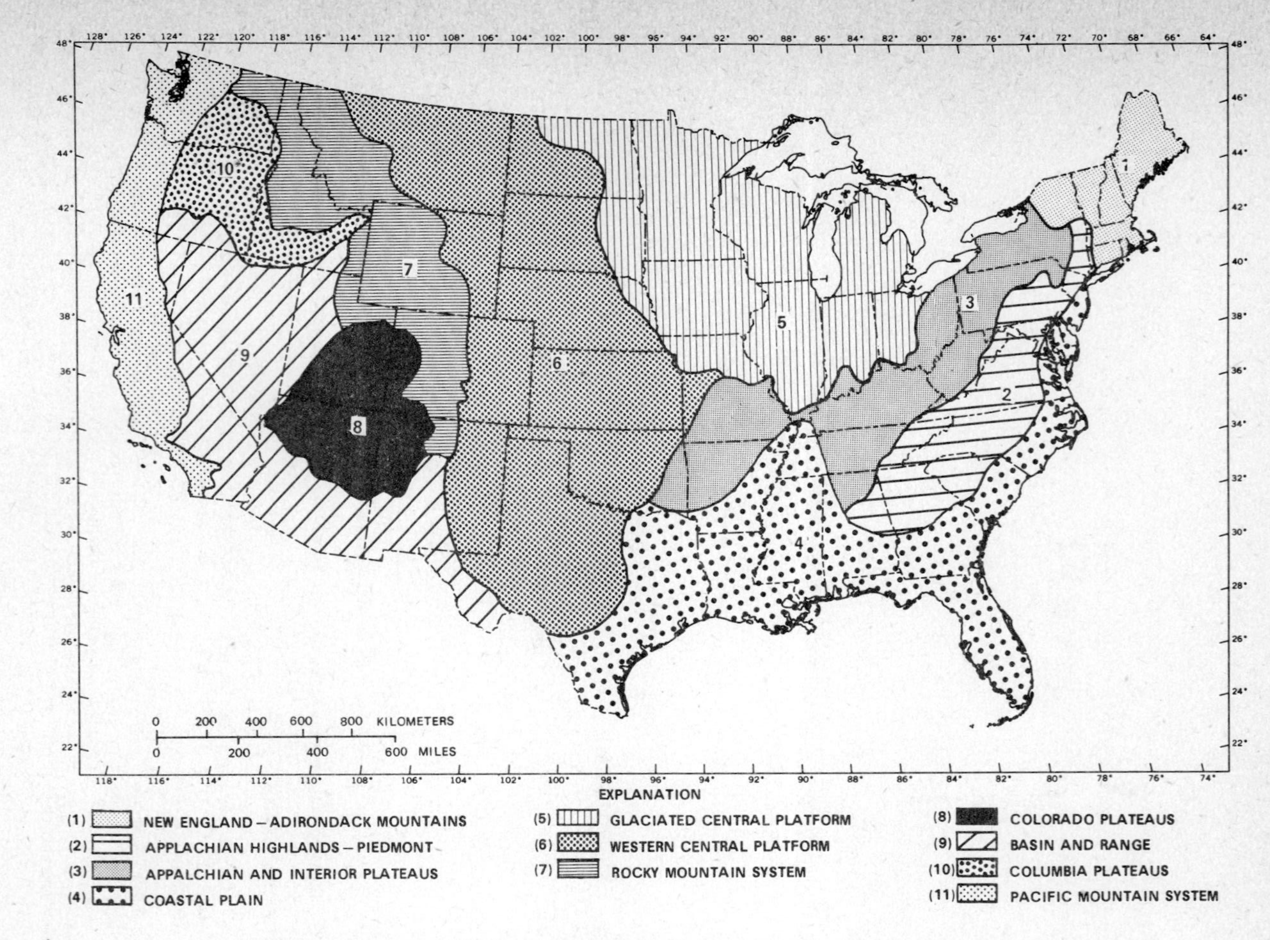

Figure 1.--Provinces of the conterminous United States selected for screening.

INTRODUCTION

Investigations to search for sites suitable for radioactive waste repositories have been underway for many years in several countries. Typically, they have been focused on a specific type of host rock. Many early studies in the United States concentrated on salt as the host rock. Subsequent studies concentrated on basalt and tuff as the host rock. More recently, crystalline igneous and metamorphic rocks have been the subject of search for waste repository sites. Only recently, however, has the emphasis, in the United States, been placed on evaluation of the earth-science questions involved in geologic disposal of high-level waste and the need to consider, at the earliest stage in the search, environmental factors that can provide multiple barriers to radionuclide migration [1, 2]. While the host rock remains important, it is recognized that several types of rock can serve as suitable host rocks. The host rock needs to be considered in conjunction with the natural hydrologic and geologic barriers afforded by the environment.

The term multiple barriers includes man-made barriers and natural barriers in the form of specific hydrodynamic, geochemical, and geologic characteristics that would impede radionuclide transport. The factors of foremost significance include a geologically stable environment, a host rock with minimal permeability and large sorptive capacity in a flow system with slow ground-water velocity, and long flow paths downgradient from the repository to the accessible environment.

The Earth Science Technical Plan Working Group [3] identified technical questions that need to be answered to achieve geologic disposal of high-level radioactive waste. A key question to which the U.S. Geological Survey program discussed here is related to the identification and characterization of suitable geologic and hydrologic environments for repositories. This question was considered in detail by a subgroup of the Earth Science Technical Plan Working Group. The recommended plan for study by this group [4] is the guiding document for the Survey's study [5].

The underlying goal of radioactive waste isolation is to prevent migration of radionuclides to the accessible environment in concentrations that may be unacceptably hazardous to humans. Accessible environment, as used here, means those parts of the environment directly in contact with or readily available for use by people. It includes the earth's atmosphere, the land surface, surface waters, and the oceans. It also includes presently used aquifers containing potable water [6]. Both man-made and natural barriers are considered in design and evaluation of a repository. The U.S. Geological Survey program is designed to evaluate geologic and hydrologic factors of a site on the basis of the natural geohydrologic system's capacity to provide multiple natural barriers for the isolation of high-level radioactive waste beyond the man-made barriers in the repository. Containment of waste by man-made barriers is not within the scope of the Survey study. However, the Survey's plan of study recognizes that engineering design and feasibility depend upon the geologic and hydrologic conditions; therefore, engineering design imposes certain requirements on geologic and hydrologic conditions at the repository site. The waste form also can be tailored to some extent to be compatible with geologic and hydrologic conditions. Specifically, the thermal load can be reduced substantially by allowing the radionuclides to decay in surface storage for several decades or by engineering design of the repository to distribute the heat load over a larger area.

This report is intended: (1) To give a brief outline of the U.S. Geological Survey's plan for evaluating the suitability of geologic environments for waste isolation; (2) to briefly summarize the salient features of the screening criteria; and (3) to describe the more important factors to be considered and their treatment in the initial step of screening Province-sized (2,600 to 260,000 square kilometers or 10^5 to 10^6 square-miles) land units.

NATIONAL SCREENING PLAN

The U.S. Geological Survey's study is designed to screen the conterminous United States to identify locations with favorable geohydrologic conditions for repository sites. The screening process consists initially of identifying 11 Provinces in the conterminous United States (fig. 1) as the land unit for initial characterization and evaluation. The Province is successively divided into smaller land units, called Regions (2,600 square kilometers to 260,000 square kilometers or 10^3 to 10^5 square miles), which are successively divided into Areas (260 square kilometers to 2,600 square kilometers or 10^2 to 10^3 square miles), and potential Sites (26 square kilometers or 10 square miles). At each stage, the process involves geologic and hydrologic characterization and evaluation of the land units with repect to screening guidelines pertinent to radioactive-waste isolation.

The screening process will be accomplished by the U.S. Geological Survey in conjunction with and a Province Working Group. Each State in the Province will be invited to participate with the U.S. Geological Survey in the Province Working Group. The Province Working Group will be comprised of earth scientists from the U.S. Geological Survey and the States in the Province. The earth scientists from the respective States who will serve on the Province Working Group will be selected from State Geological Surveys or other earth-science oriented agencies, universities, or the private sector.

Geologic and hydrologic factors will be the sole basis for characterization and evaluation of Regions and Areas by the Province Working Group. Nongeologic factors will be the subject of a study conducted under the sponsorship of the U.S. Department of Energy. Nongeologic factors such as land use, accessibility, population, and institutional factors will be considered by the U.S. Department of Energy in consultation with the States in recommending Regions for study following Province evaluation and in selecting Areas for study following Region evaluation.

The characterization of the Province will be based on data and information available in published reports and unpublished data in the files of Federal agencies, State agencies and the private sector. Geologic and hydrologic characterization of the Province is guided to a large extent by the nature of the criteria for containment of high-level radioactive waste.

Regional subdivisions of the Province will be defined on the basis of ground-water flow systems, similarity of geologic features, and the detailed evaluation of the Province made in applying the screening guidelines. The evaluation of the Province will result in a threefold subdivision; (1) those Regions that appear to contain favorable geohydrologic environments for containment of high-level radioactive waste; (2) those Regions that appear unfavorable; and (3) those Regions where data are insufficient to judge suitability of the environment for waste containment.

Regions that appear to contain favorable environments will be selected for characterization. Regional characterization will involve onsite reconnaissances, laboratory analyses, hydrologic modeling, and the interpretation of geologic, hydrologic, and geophysical data. Following characterization, the geologic and hydrologic properties of Regions will be evaluated with repsect to their conditions for containing waste and Areas will be selected for detailed study. Characterization of Areas will involve more extensive onsite work, including such activities as test drilling, hydrologic testing, borehole and areal geophysical surveys, and other onsite and laboratory work. Evaluation of Areas would lead to the identification of potential Sites for detailed characterization.

PROVINCE SCREENING CRITERIA

It generally is believed that repositories in geologic media can provide acceptable isolation of radioactive waste. Wastes from a deep repository could be exposed to the biosphere either (1) by some geologic process such as tectonism, diapirism, or erosion that directly exposes the waste or (2) by some process that transports the wastes to the biosphere. Geologic criteria stress suitability of host rock and geologic stability of the site. Hydrologic criteria stress slow transport of radionuclides, long flow paths, and retardation of radionuclides [6, 7, 8]. Transport by ground water is acknowledged to be the most probable mechanism for moving radioactive waste from the repository to the accessible environment. Knowledge of the ground-water flow path is thus important in providing an isolation system if waste should leak from the repository.

The essence of the evaluation approach used for assessing the suitability of geohydrologic environments for waste isolation is the interrelation of individual factors and their mutual interaction in the system. Initially, however, the factors are evaluated individually. Generalizations can be made concerning favorable characteristics for each factor. Certain minimum characteristics for individual factors, such as thickness of host rock, are referred to as criteria. Criteria are useful as guidelines to preliminary evaluation of suitability, but it is unlikely and unnecessary that a site be ideal with respect to all criteria. The ultimate judgment of suitability will be on the basis of the capability of the system as a whole to isolate the waste.

Screening criteria may be envisaged to evolve with increasing specificity from Province to Region to Area to Site. Criteria on thickness of host rock, for example, might specify a greater thickness at the Area screening stage; criteria on rates of erosion, a factor that varies widely throughout large areas, probably would not be considered during Province or Region screening.

The following factors, modified from the report of Subgroup I of the Earth Science Technical Plan [4], will be considered in establishing criteria for the Province stage of screening. The criteria established need to be consistent with the criteria being formulated by the U.S. Nuclear Regulatory Commission [6] in regulating the disposal of high-level radioactive waste by the U.S. Department of Energy.

Repository host rock. The area needs to be underlain at least in part by a system of rocks containing one or more suitable host-rock units. The following factors need to be considered:

Mineability - It should be possible to excavate the repository using available mining methods and technology.

Thermal conductivity - The thermal conductivity of the rock unit should be high enough to accommodate the thermal stress imposed by the particular waste form without causing significant increase in permeability or detrimental alteration of the waste form.

Fractures - The rock unit should have a minimum of natural and induced permeable fractures.

Permeability - The rock unit should have a minimal permeability.

Dimensions and geometry - The rock unit should be sufficiently thick and extensive in area to accommodate the facility.

Depth - The repository host rock should occur at sufficient depth to minimize the possibility of exposure through geomorphic processes, tectonic processes, or both.

Homogeneity - The rock unit should be sufficiently homogeneous to make it possible to predict its long-term physical properties in advance of mining and development.

Sorption capacity - The rock unit should have a high radionuclide sorption capacity to enhance radionuclide residence time in the flow system.

Geochemical properties - The geochemical properties of the rock unit and contained water should prevent or minimize chemical reactions with the wastes that would facilitate the transport of radionculides from repository sites.

Thickness - The thickness of the host rock must be great enough to contain the repository, and provide an additional thickness to impede movement of ground-water flow above and below the repository.

Lateral extent - The host rock must extend laterally beyond the repository.

Ground-water flow system. The essential attribute of the flow system is to provide long residence time before the water enters the accessible environment.

Travel time - The rocks in the system between the repository site and the discharge area should have minimal permeability, a long flow path, and small hydraulic gradients to provide a long residence time.

Flow direction - Ground water in a substantial part of the area should have a significant downward or lateral component of flow. There should be no upward flow, particularly if the area contains numerous oil, gas, or other exploratory holes or a large potential for such holes being drilled in the future.

Uniformity - The hydraulic characteristic of the system should be sufficiently uniform to permit spatial extrapolation of these characteristics to the nearest discharge area. In general, it is preferable that rocks along the flow path should have granular rather than fracture porosity.

Sorption - Rocks with high sorptive capacity should dominate along the ground-water flow paths.

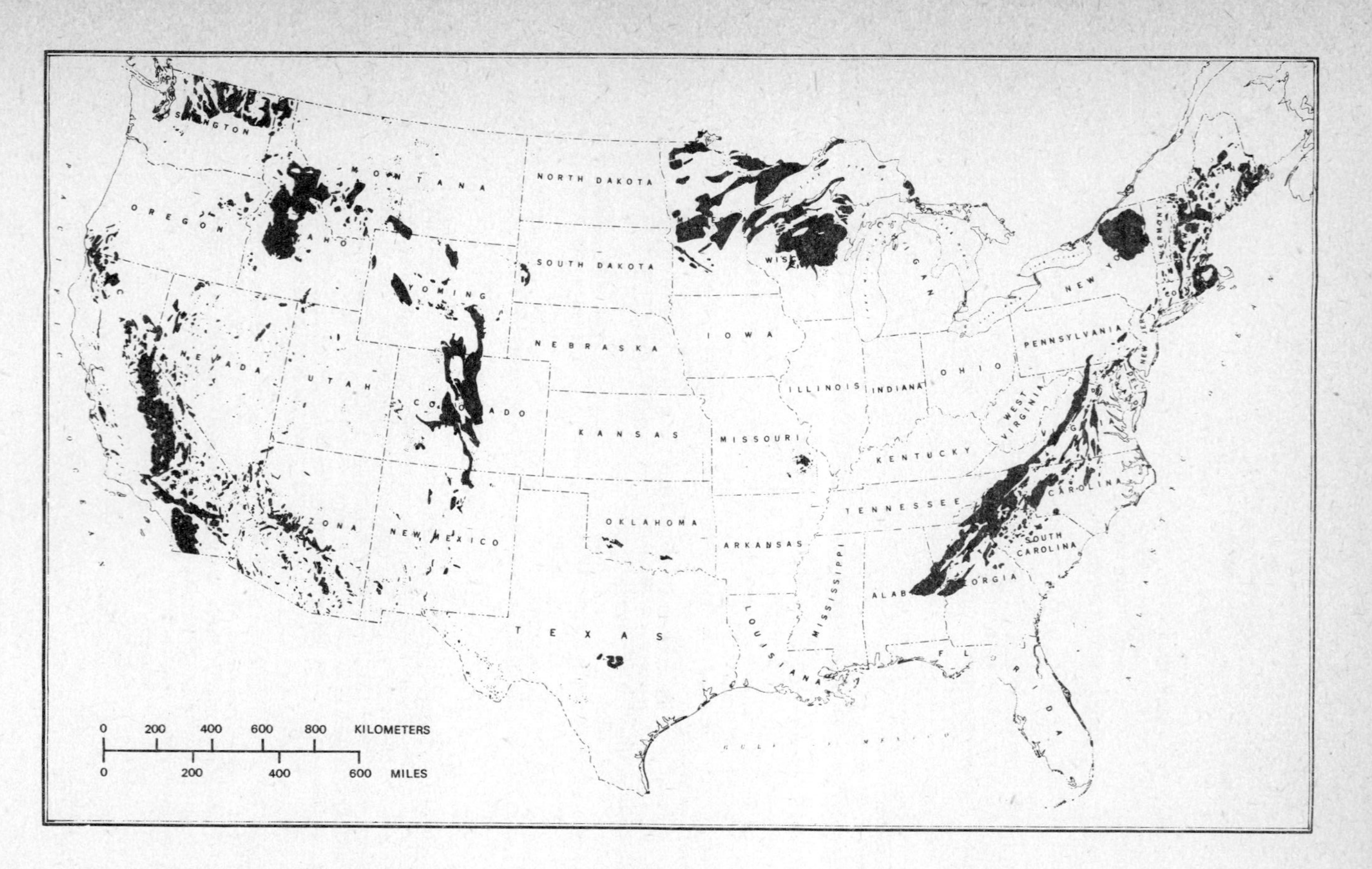

Figure 2.--Distribution of exposed crystalline igneous rocks (black) in the conterminous United States (modified from Bayley and Muehlberger [9]).

Water quality - To minimize the possibility of future intrusion of the repository by water-well drilling, the potential host rock unit should be underlain by, and, at least immediately overlain by, nonpotable water. An aquifer system containing potable water near the land surface would minimize the incentive to drill deeper in search of water.

Tectonic conditions. Certain geologic processes resulting from tectonic activity could disrupt the repository environment and facilitate the mobilization of waste radionuclides in ground water, or possibly even directly expose the wastes. The criteria need to consider the following areal factors in terms of their effects on potential repository environments: (1) known active faults, (2) high seismic intensity, (3) recent volcanic activity, and (4) persistent uplift.

Mineral resources. The intent is: (1) To avoid mineralized zones at depths greater than the potential host rocks to minimize the possibility of radionuclides escaping from the repository through preexisting boreholes that could not be sealed satisfactorily; and (2) to minimize the potential of penetrating the repository in the future by holes drilled for mineral exploration or development. In this sense, aquifers containing potable water are considered a mineral resource.

GEOLOGIC AND HYDROLOGIC CHARACTERIZATION

Screening criteria do not apply equally at each stage of screening. Processes such as erosion and factors such as topography vary locally and criteria related to them generally are not applicable to the screening process for a large area, for example, a Province. In the strategy presented here nongeologic factors are not considered until after the Province stage, in order that a comprehensive characterization of the geology and hydrology of the Province can be developed.

Host rock and ground-water flow systems are the two most significant factors in Province characterization. The significance of the presence of the host rock is obvious; the screening process takes place only where favorable host rocks occur. Ultimately, the ground-water flow system, is most important though its significance is not as obvious.

Host Rocks

Granite and other silicic to intermediate crystalline plutonic rocks, argillaceous rocks, salt deposits (both bedded and dome salt), welded ash-flow tuffs, and basaltic lava flows have been considered as potential host media for deep repositories. In the conterminous United States these rocks are widely and unevenly distributed. General distribution of those potentially favorable host media will be discussed in this section and are briefly noted in table I. Other factors important to consider for host rocks are their extent and uniformity at depth, their surface and subsurface structure, and their geologic history.

Granites

Granites and related plutonic igneous rocks are widely distributed throughout the United States (fig. 2) [9, 10]. These rocks include all holocrystalline medium-grained to porphyritic rocks ranging in composition from granite to diorite. Most of the exposed rocks of this type in the United States are granites and granodiorites. Quartz-deficient rocks, such as diorite and syenite, are much less abundant; nevertheless, they may have potential as host rocks. Most crystalline rocks being considered have an areal extent of hundreds to thousands of square kilometers and extend thousands of kilometers deep. These rocks clearly have sufficient lateral extent and depth to hold repositories. Except for a few of the plutonic bodies, only the basic petrochemistry is known. Many plutons are known to have chemical and mineralogical zonations. The zonations at depth in virtually every plutonic body in the United States are unknown. Additionally, very little is known of the faulting, fracturing, jointing, and other structural details of plutonic rocks at depth. This information can only be extrapolated from surface and near surface data and from a few deep mines in the United States and elsewhere.

Argillaceous Rocks

Thick argillaceous rocks may have great potential for subsurface emplacement of radioactive waste. Their minimal permeability, relatively large plasticity, large ion-exchange capacity, and widespread distribution are desirable character-

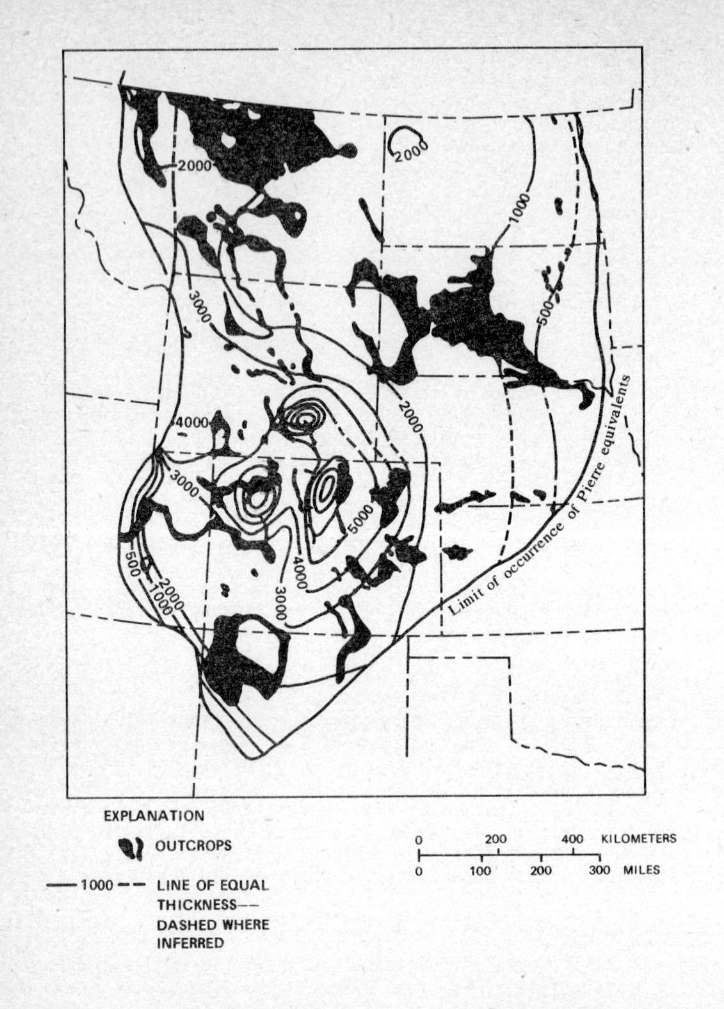

Figure 3.--Distribution and thickness of Pierre Shale and equivalent rocks in the western interior of the United States (modified from Tourtelot [15]).

istics. However, many argillaceous rocks include mineral constituents that have undesirable properties. Some shales, mudstones, and claystones contain organic matter, montmorillonite, and other so-called expanding clays that start to alter at temperatures as low as 70° Celsius.

Sedimentary rocks underlie the majority of the land area of the United States and about one-half are shale. The porosity of the average clay is 27 percent and of the average shale is 13 percent [11]. The average composition of shale, determined from the analyses of 10,000 samples [12] is 59 percent clay minerals. The common clay minerals in shale are illite, chlorite mica, montmorillonite, and kaolinite [13]. Illite is probably the dominant clay mineral. Montmorillonite is abundant in Mesozoic and Cenozoic sedimentary rocks in the United States but rare in pre-Mesozoic rocks.

Shales of marine origin traditionally have received more attention as host rocks than nonmarine shales, because marine shales generally are more uniform in composition. Nonmarine shales commonly contain coarser-grained and more permeable beds. Shales in areas of abundant faults and fractures commonly have greater secondary permeability and their characteristics at depth are less predictable. Occurrences of shales and other argillaceous rocks of thicknesses in excess of 150 meters (500 feet) are discussed briefly and modified largely from Merewether and others [14].

In parts of California, Oregon, and Washington of the Pacific Mountain Province, Tertiary shales are as thick as 3,000 meters (10,000 feet) thick. Some are montmorillonitic; others are in areas of complex structure. Thick Paleozoic shale formations in parts of California, Nevada, Utah, and New Mexico of the Basin and Range Province, have undergone mild to intense structural deformation. However, sections of the Mancos Shale of Cretaceous age as thick as 1,500 meters (5,000 feet), but with significant montmorillonite content, occur in Utah, Arizona, New Mexico, and eastward to the Rocky Mountain System. In the Rocky Mountain System, Colorado Plateaus, and Western Central Platform Provinces, thick Paleozoic formations consisting mainly of shale crop out in Montana, Idaho, Wyoming, Utah, Colorado, and New Mexico. Cambrian, Mississippian, and Pennsylvanian shales are as thick as 600 meters (2,000 feet). Thick sequences of Cretaceous rocks include the Colorado and Graneros Shales, the Benton Formation, and the Mancos, Carlile, and Pierre Shale (and Pierre equivalents, the Clagett, Bearpaw, and Lewis Shales). The Pierre and equivalent rocks are widely distributed (fig. 3 [15]) and consist of claystone, shale, mudstone, and many thin beds of bentonite. The Pierre ranges in thickness from less than 150 meters (500 feet) in the eastern Dakotas to more than 1,500 meters (5,000 feet) in southeastern Wyoming and central Colorado. The Pierre is largely devoid of aquifers, and, although highly porous locally, interstitial permeability is very small. Clay minerals in the Pierre consist of 25 to 45 percent montmorillonite, 35 to 45 percent mixed-layer illite-montmorillonite, 15 to 25 percent illite, and about 5 percent each of kaolinite and chlorite. In much of the Glaciated Central Platform Province, Paleozoic shale, mudstone, and claystone generally do not contain much montmorillonite but they are less than 150 meters (500 feet) thick. In scattered structural basins, however, the clay-rich rocks are much thicker, although many of the best sequences are at great depths, intensely faulted or folded, or contain combustible hydrocarbons. The Michigan basin of southern Michigan, northern Indiana, and northwestern Ohio has thick Devonian and Mississippian shale at moderate depth. The Coastal Plain Province is underlain by thick, predominantly clay strata of Mesozoic and Cenozoic age. Much of the shale has high porosity and is poorly indurated. Clay sequences in the Gulf Coast and Mississippi Embayment contain a large percentage of montmorillonite. The shales are interbedded with aquifers that contain saline water at depth.

The Appalachian Highland's Piedmont Province contains thick shales of Paleozoic age, which, locally, are in areas of very complex structure. The most promising host rocks are of Ordovician, Devonian, and Mississippian age. In eastern New York, northwestern New Jersey, and eastern Pennsylvania, Ordovician shales are 600 to 2,400 meters (2,000 to 8,000 feet) thick. Devonian shales crop out on the north and east flanks of the Appalachian basin in New York, Pennsylvania, Maryland, West Virginia, Virginia, and Tennessee, and on the west flank of the basin in Kentucky and Ohio. The thicker, less-deformed sequences of shale occur in localities on the north and west flanks of the Appalachian Highlands.

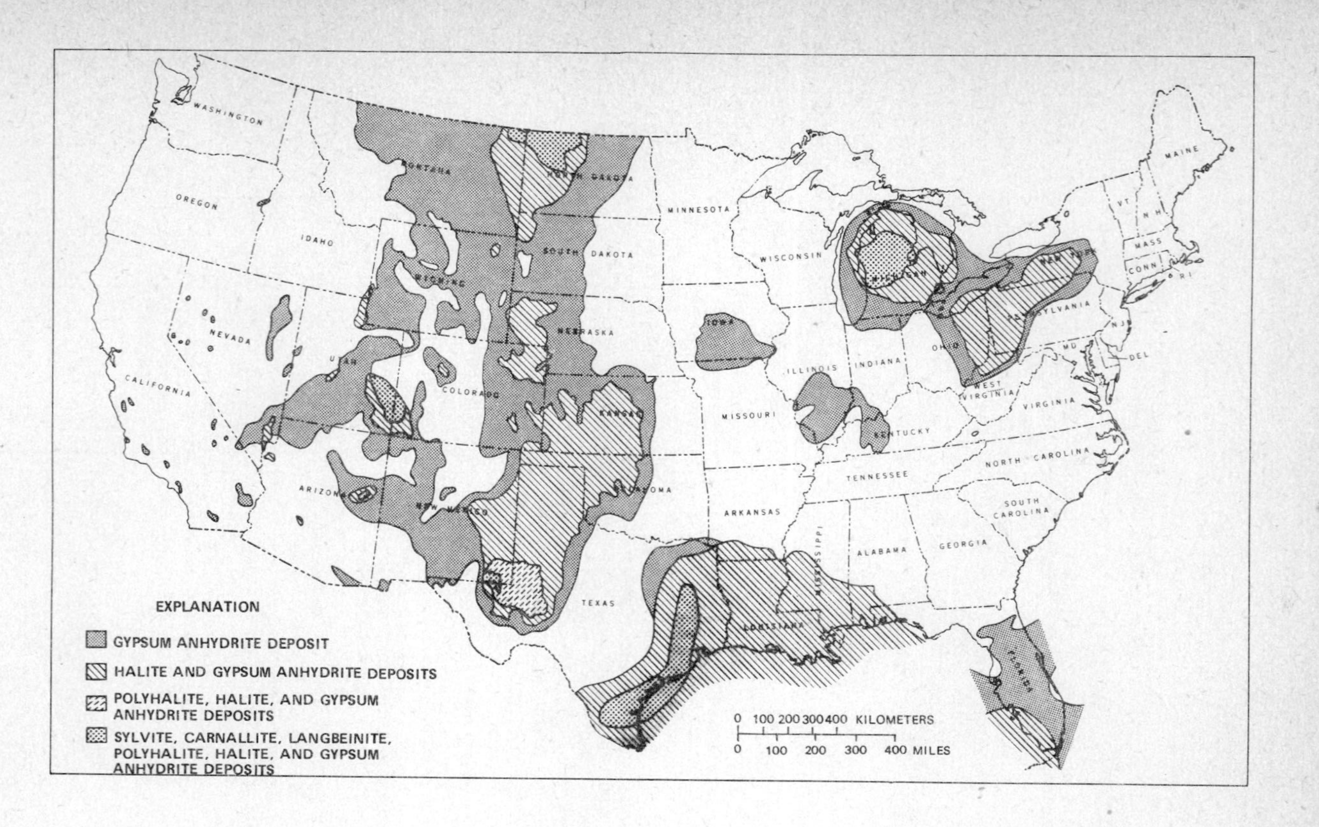

Figure 4.--Marine evaporite deposits of the United States (modified from Smith and others [16]).

Salt Deposits

Rock-salt (principally halite) deposits in the United States are widely distributed (fig. 4) [16]. Some deposits are extensive such as the Silurian salt that underlies most of Michigan and large parts of New York, Ohio, Pennsylvania, and West Virginia. Salt deposits range in age from Silurian to Pliocene(?).

Salt has received much attention as a repository host rock and exploration has been directed at bedded deposits in the Permian basin of New Mexico and Texas, the Paradox Basin of Utah, and salt domes of the Gulf Coast area with less extensive programs in the Michigan and Salina Basins. An extensive compilation of the properties of salt deposits relative to high-level radioactive waste disposal was provided by Johnson and Gonzales [17].

Basalt and Welded Ash-Flow Tuff

Under favorable conditions thick, massive, basalt and welded ash-flow tuff may be suitable as host rocks. Investigations for repository siting in thick Tertiary flood basalts of the Columbia Plateau are being conducted in southern Washington at the Hanford Reservation, and in welded tuff at the Nevada Test Site in southern Nevada. The most widespread and possibly the most suitable basalts and welded tuffs for repositories are of Cenozoic age in the western United States.

Some lower Tertiary (Eocene and Oligocene) volcanic fields, such as the San Juan field of southwestern Colorado, the Challis Volcanics of south-central Idaho, and the Lowland Creek Volcanics of western Montana, contain thick sequences of ash-flow tuffs, some of which may be thick and persistent at depth. In general, the lower Tertiary rocks are predominantly intermediate in composition (54 to 65 percent silica) and tend to be complexly intercalated lavas, breccias, mudflows, and bedded ash, all of which are likely to have unpredictable lateral extent.

Upper Tertiary (Miocene and Pliocene) volcanics (fig. 5) [18] form a large and complex assemblage in the western United States. Some of the more extensive volcanic fields will be discussed briefly.

Upper Tertiary andesites of the Cascade Ranges of Washington, Oregon, and California locally contain thick olivine basaltic and andesitic lavas with only minor amounts of ash-flow tuff. Here, as in some lower Tertiary volcanic fields, the sequences of flows are likely to be composite and complex with relatively unpredictable lateral extent.

The Columbia Plateau in southeastern Washington, northeastern Oregon, and west-central Idaho is covered by the Columbia River Basalt Group of Miocene to Pliocene age. These form a terrane unique in the United States. Throughout most of the plateau, the basalts are flat-lying and virtually undeformed, but along their west margin in Washington they are deformed into steep-sided, west-northwest-trending anticlines as a result of deformation during the Pliocene. The Columbia River basalts cover an area of 260,000 square kilometers (100,000 square miles) and near their center of distribution, they exceed 3,000 meters (10,000 feet) in thickness. Individual flows average 20 to 60 meters (75 to 200 feet), but some are more than 120 meters (400 feet) thick. The Columbia River Group basalts are monotonously uniform tholeiites that differ little in composition through 900-meter (3,000-foot) sections.

Although volcanic rocks underlie much of southeastern Oregon and northeastern California, thin 6 meters (20 feet or less) alkali olivine, basalt flows, of Miocene and Pliocene age locally accumulated to thicknesses exceeding 1,500 meters (5,000 feet). Notable examples of these rocks such as the Steens Basalt in Oregon and the so-called Warner Basalt in California are overlain by a complex assemblage of Pliocene basaltic flows and volcaniclastic rocks. The total sequence is very broken by Basin and Range faulting.

The Idavada Volcanics of southwestern Idaho are as much as 900 meters (3,000 feet) thick and consist of welded ash-flow tuffs and rarer silicic latite and rhyolitic lavas of Miocene age. Overlying these thick tuffs is the Banbury Basalt, also of Miocene age, which thickens northward from the Nevada border to about 300

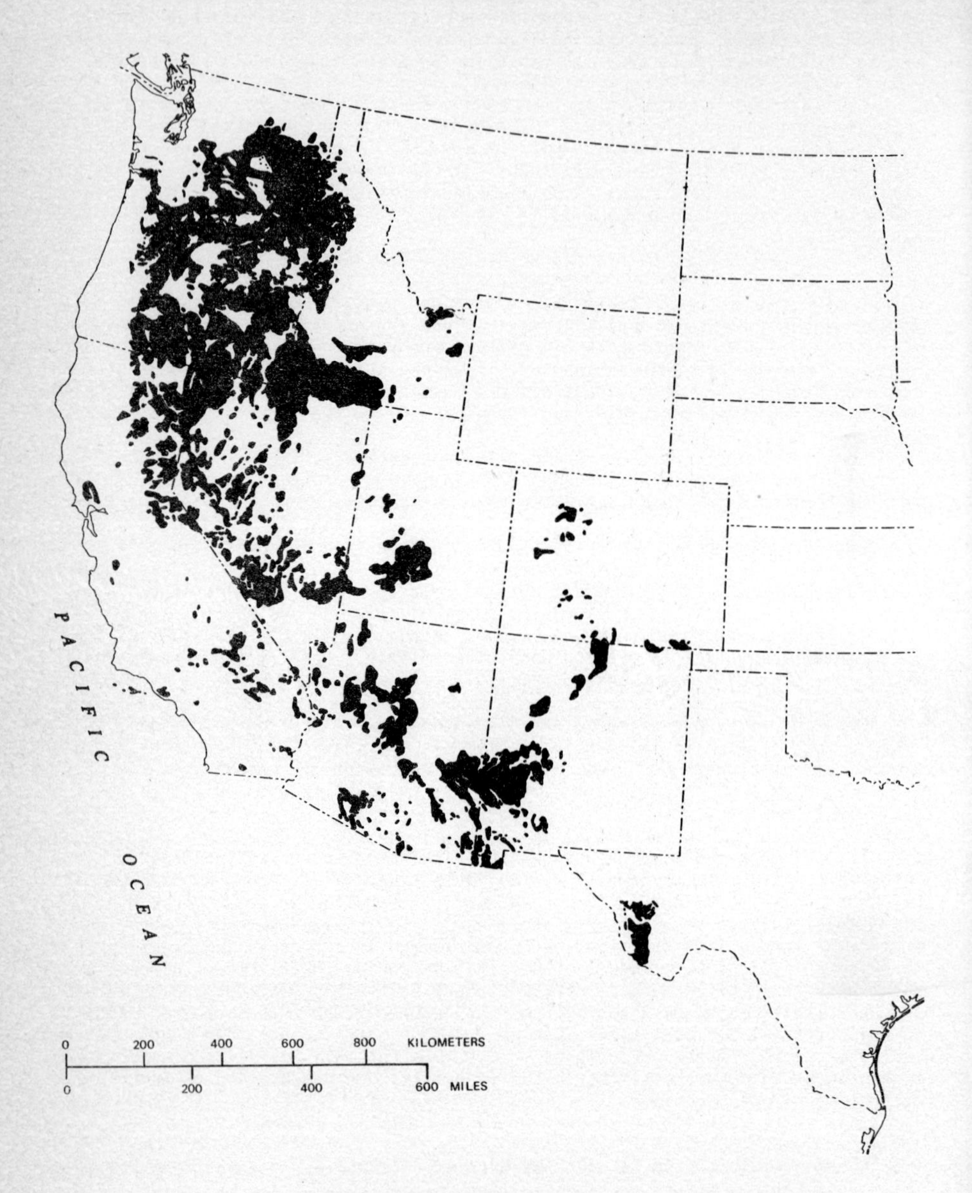

Figure 5.--Distribution of Miocene and Pliocene volcanic rocks (black) in the western United States (modified from King and Beikman [18]).

meters (900 feet). Overlying volcanic rocks of the Snake River Plain are of Quaternary age.

Miocene and Pliocene volcanic rocks form an extensive and complex assemblage in the Great Basin of Nevada, and extend into Utah and California. In central and western Nevada, significant volumes of rhyolitic ash-flow tuffs were extruded between 34 and 20 million years ago. These Oligocene and Miocene flows are typically more than 450 meters (1,500 feet) thick and consist of a succession of ash-flows or cooling units. They are widely traceable from range to range. Calderas and nearby accumulations of these flows likely will contain very thick sections of welded tuff.

In western Nevada, most of the volcanic rocks belong to the Pliocene sequence. They were erupted from local centers where they may attain thicknesses of several thousand feet.

A different sequence of upper Tertiary volcanic rocks occurs in southern Nevada. Here an extensive and nearly unbroken field centering in the Nevada Test Site of southern Nye County attains a thickness of as much as 12,000 meters (40,000 feet). Older tuffs crop out around the edges of the volcanic field, but the most extensive surface rocks are Miocene and Pliocene ash-flow tuffs erupted from the Timber Mountain and Black Mountain calderas. Several older calderas have been covered by younger volcanic rocks and are known only from drilling and geophysical surveys.

Locally, additional volcanic fields may contain thick uncomplicated basalts, ash-flow tuffs, or both; these include the Marysvale volcanic area in west-central Utah, the San Francisco Peaks in northern Arizona, and other isolated areas such as the Galiuro and Chiricahua Mountains of south-central Arizona.

Other Potential Host Rocks

Anhydrite, rocks in the unsaturated zone, carbonate rocks, metamorphic rocks, and mafic plutonic rocks may be suitable host rocks for some types of radioactive or toxic wastes with an appropriately engineered waste package and repository.

The suitability of anhydrite for waste has yet to be extensively tested. This rock type is found commonly with limestones and shales associated with halite and gypsum in virtually all marine evaporite basins (fig. 4 and [19]).

The use of rocks in the unsaturated zone depends on the hydrologic environment to provide a major barrier to radionuclide transport [20]. Areas where unsaturated rocks are thick enough and deep enough to accommodate high-level or transuranic contaminated waste may be quite restricted. Deep alluvial basins in the Basin and Range Province are the most likely location to find thick unsaturated sections of alluvium, however, the thermal conductivity of dry alluvium and other unsaturated rocks is small.

Normally, limestones are readily soluble, therefore they are not thought of as potential host rocks. However, thick limestone lenses surrounded by low permeability, sorptive shales, or blocks isolated hydrologically by tight faults or intrusive barriers may be suitable. The anisotropy of schist and gneiss would add uncertainty to predictive models for these media; the sorptive properties of quartzite do not appear promising. Metamorphosed equivalents of clay-rich rocks, such as argillite and metasiltstone, need to be given consideration as host rocks, and may, because of their greater induration, be better suited for mining than the less competent shales and siltstones.

Mafic plutonic rocks may be suitable as host rocks for certain waste packages. Their higher temperature of formation and their lower silica, potassium, and sodium content may enhance their suitability for containing some wastes.

Ground-Water Flow Systems

The nature of major flow systems in the Provinces of the conterminous United States is briefly described in Table I. Definition of flow systems is considered the basic step in hydrologic characterization of the Province. Flow system boundaries are the basis for separation of the Province into logical subunits, Regions,

Table I.--Summary of potential host rocks, geologic framework, and nature of ground-water flow systems in the Provinces of the conterminous United States (fig. 1)

	Province	Principal Potential Host Rocks	Geologic Framework	Nature of Ground-Water Flow Systems
1.	New England-Adirondack Mountains	Crystalline Igneous rocks	New England--complexly faulted elongate fold belts of Precambrian metamorphic rocks and Paleozoic metasedimentary rocks intruded by large masses of Paleozoic granite. Adirondacks--Precambrian mountain mass composed of marble and schist intruded by Archean granite and syenite and later by Precambrian anorthosite and gabbro.	Flow and head in bedrock of low permeability and overlying glacial aquifers greatly influenced by local topography and surface-water features.
2.	Appalachian Highlands-Piedmont	Crystalline Igneous rocks, Shales	Appalachian Highlands--Elongate mountainous anticlinorial belt of Precambrian granites and metamorphics thrust westward over Paleozoic rocks. Piedmont--nonmountainous belt of highly complex early Precambrian gneisses and intrusives and lower Paleozoic metasediments intruded by later Paleozoic granitic plutons.	Flow and head in Precambrian and granite bedrock of low permeability, largely controlled by topography and surface water features; folded limestone of Paleozoic age locally cavernous and highly permeable at shallow depths supporting large springs; sandstone aquifers of moderate extent and permeability; flow systems generally related to local recharge in interstream areas and discharge to surface water features; thermal springs along western edge of Province indicative of moderately deep (1 to 1.5 km) convection in artesian system.
3.	Appalachian and Interior Plateaus	Shales, Evaporites, Granites	Appalachian and Interior Plateaus consist of gently dipping, gently folded, domed Paleozoic sandstones, shales, carbonates, and evaporites. In southern Missouri exposing Precambrian crystalline rocks; Paleozoic shales, sandstones, cherts, and carbonates in southern Arkansas intensely folded and faulted.	Regional flow in low to moderately permeable sandstones and carbonates; carbonates locally of high permeability at shallow depth due to fractures and solution channels support large springs.

Table I.--Summary of potential host rocks, geologic framework, and nature of ground-water flow systems in the Provinces of the conterminous United States (fig. 1)--Continued

4.	Coastal Plain	Clays, Shales, Evaporites (salt domes and bedded salt)	Seaward dipping thickening wedge of Mesozoic and Cenozoic sandstones and shales with some evaporites and limestones; underlain by a basement of Paleozoic and Precambrian rocks.	Regional flow in sand and limestone aquifers with intervening clay confining layers; predominant flow direction seaward; discharge upward through confining layers and to streams; geopressured zones in northern Gulf coast of Texas and Louisiana.
5.	Glaciated Central Platform	Crystalline Igneous rocks Evaporites and Shales	Precambrian igneous and metamorphic shield rocks on the northwest overlain by nearly horizontal Paleozoic sandstones, carbonates, shales, and evaporites; deep basin deposits in Michigan and Illinois; broad domal structures.	Regional flow in Paleozoic sandstone and carbonate aquifers; highly mineralized water at depth in basins; glacial aquifers locally overlie bedrock.
6.	Western Central Platform	Shales and Evaporites	Horizontal to gently dipping broad syncline of Paleozoic through Mesozoic sandstones, shales, carbonates, and evaporites; deep sedimentary basins and structural highs. Capped with Tertiary sands and gravels.	Regional flow in layered sandstone and carbonate aquifers; thick confining beds of shale; deep basins contain highly saline water (300,000 mgl) and geopressured zones; predominant flow direction toward the east from upturned beds cropping out along the eastern front of the Rocky Mountains. Underpressured zones locally in upper part of section and at depth near recharge areas near mountain front. Extensive fluvial deposit aquifers from Nebraska south into Texas and glacial aquifers in North Dakota and South Dakota overlie Tertiary and older rocks.
7.	Rocky Mountain System	Crystalline Igneous and Metamorphic rocks; Shales and Volcanic rocks	Igneous and metamorphic folded and faulted core rocks of Rocky Mountains with intermontane basins of Paleozoic through Cenozoic shales, carbonates, evaporites, and sandstones. Tertiary intrusive and volcanic rocks.	Regional flow in layered sandstone and carbonate aquifers with shale confining beds in intermontane basins; local recharge and discharge controlled by topography and surface water features in fractured igneous and metamorphic rocks; hydrothermal convection systems related to magmatic sources at depth.

Table I.--Summary of potential host rocks, geologic framework, and nature of ground-water flow systems in the Provinces of the conterminous United States (fig. 1)--Continued

8.	Colorado Plateaus	Shales, Evaporites, and Volcanic rocks	Flat-lying to gently warped layers of Paleozoic, Mesozoic, and Cenozoic sandstones, shales, limestones and evaporites with Tertiary volcanic rocks; broad basins, long monoclinal flexures and dome mountains.	Regional flow in layered sedimentary rocks; chief aquifers are sandstones and carbonates; discharge to major streams; highly saline water at depth in deep basins and in association with salt beds; under-pressured zones reported in deep sedimentary basin of New Mexico.
9.	Basin and Range	Volcanic rocks, Crystalline igneous and Metamorphic rocks, unsat-urated alluvium, Shales	Elongate blocky mountains of faulted Precambrian, Paleozoic Mesozoic, and Cenozoic rocks; deep alluvium-filled intermontane basins; intrusive igneous stocks and plugs of Mesozoic and Tertiary age; extrusive ash-flow tuffs, rhyolites, and basalts of Cenozoic age.	Flow within closed basins; interbasin flow between closed topographic basins through permeable bedrock; interbasin flow in alluvial channels between basins with integrated surface drainage; deep regional flow systems in carbonate and volcanic rocks including hydrothermal convection flow systems.
10.	Columbia Plateaus	Basalts	Regional shallow structural basin of Cenozoic basaltic lava flows; locally faulted and folded; mountain range on the west consisting of elongate chain of andesitic volcanic cones of Pliocene and Quaternary age.	Basaltic lava flows range from highly permeable to dense nearly impermeable creating regional aquifers with hydraulic and hydrothermal convection systems and perched aquifers separated by confining beds. The ground water principally discharges to the major streams; local discharge to a few closed basins.
11.	Pacific Mountain System	Crystalline Igneous rocks, Tuffs, and Basalts	Consists of several complex elements: Large uplifted and tilted blocks of Jurassic granite with inliers of Paleozoic and Mesozoic metasediments; folded and faulted Tertiary and Mesozoic sedimentary rocks; deep elongate troughs filled with Tertiary and Quaternary sediments.	Regional flow in deep intermontane sedimentary basins with locally overpressured zones at depth; hydro-thermal convection flow systems, igneous and meta-morphic rocks of low permeability support shallow local flow systems related to topography and surface drainage.

which can be analyzed by a systems approach. This early recognition of the flow systems will maintain a framework within which the areal scope of study is clearly established from Region through Potential Site characterization. While the Site will, and the Area may, encompass a smaller tract than the flow system, the analysis of ground-water flow and radionuclide transport from the repository will necessitate some knowledge of characteristics and flow distribution in the entire flow system.

Data useful in delineating flow systems include potentiometric head, geologic framework, geochemistry and temperature of ground water, and water budget components. Using such data, flow systems have been delineated in Nevada and parts of Utah [21, 22, 23, 24, 25]. These data are used initially in delineating boundaries of ground-water flow systems. Though the geologic or hydrologic data bases are far from complete, flow-system boundary definition will become better defined through identification and collection of critical data.

Potentiometric Head

Potentiometric head is the single most definitive factor used in defining flow within a system and in delineating system boundaries. The interpretation of potentiometric head distribution needs to recognize the three-dimensional aspects of the factor as affected by sources (recharge) and sinks (discharge), fluid density variations, and the permeability fabric of the flow system. Ground-water divides on the surface of the zone of saturation may not extend vertically downward to the base of the flow system. For example, subsidiary flow systems at shallow depth in topographically closed basins are underlain by regional transbasin flow systems in parts of the Basin and Range Province.

Contemplated depths for mined repositories are generally from 300 to 1,000 meters (1,000 to 3,000 feet). We are concerned with defining hydrodynamic conditions existing at these depths and below in addition to shallower depths. Potentiometric heads at shallow depths generally are measured in water wells. Drill-stem tests in oil and gas test holes provide the major source of measurements of potentiometric head at depths of 300 to 900 meters (1,000 to 3,000 feet) and below. The hydrostatic pressure, p, of water increases linearly with depth below the water surface according to the relationship:

$$p = \rho_w \, g \, \ell \qquad (1)$$

where ρ_w is the density of water, g is the gravitational acceleration; and ℓ is the depth below the water surface. In fresh water, the hydrostatic gradient is 9.8 kilopascals per meter (0.43 pounds per square inch per foot).

In dynamic systems, departures from the hydrostatic pressure grade line expressed in equation 1 are common. At shallow depth, a departure from the hydrostatic pressure grade line may indicate upward or downward components of flow. Anomalously large or small departures from the hydrostatic pressure grade line indicate the action of natural mechanisms operating in the system. Hanshaw and Zen [26] summarized the several conceivable mechanisms that give rise to anomalous pressures at great depth. These include compression, loading, and phase changes, among other mechanisms. Hanshaw and Bredehoeft [27] point out that the maintenance of anomalously high fluid pressures is dependent on a continuous source of fluid and zones of low permeability to impede release of fluid. Conversely, maintenance of anomalously low fluid pressures is dependent on a continuous sink of fluid and zones of low permeability to impede inflow of fluid.

A brief review of anomalously high pressure zones in the United States is given in Wallace and others [28]. Geologic ages of formations in geopressured zones range from Paleozoic through Mesozoic to Cenozoic. Maximum fluid pressure gradients are reported to range from 16.5 to 24.0 kilopascals per meter (0.73 to 1.06 pounds per square inch per foot).

Anomalously low hydrostatic pressures at great depths, not as commonly reported in the literature are known in the Western Central Platform Province in Mississippian and Pennsylvanian rocks of southeast Colorado, in the Cretaceous and Tertiary rocks of North Dakota, South Dakota and Wyoming by C. W. Spencer [29]. Berry [30] in Hanshaw and Hill [31] reported underpressured zones in northwest New

Mexico of the Colorado Plateaus Province. Zones of anomalously low fluid pressures are of significant interest in isolation of high-level waste because the fluid head gradients are, by inference, directed downward, away from the accessible environment, and the zones are separated from the accessible environment by thick confining layers of low permeability.

Temperature

The earth's heat-flow field can produce fluid convection from the effects of water buoyancy caused by temperature differences. The coupling of fluid and heat-flow fields produces mutual effects on the resultant flow fields. In geothermal systems in areas with above normal heat flow and temperature gradients, it is to be expected that convective flow occurs by effects of both hydraulic and heat flow [32]. The discharge of hot springs is an example of such mixed convection, being driven hydraulically by the head difference between the recharge area and discharge outlet, and the density difference between cold downflowing water and hot upflowing water.

Thermal springs, having a temperature greater than the average air temperature, are surface manifestations of the existence of deep thermal convective flow systems. Thermal springs function as a discharge point for heat flow as well as ground-water flow. Consequently, the temperature and flow of a spring has significance in terms of the heat-flow system and the ground-water flow system. Inferences on the magnitude of the ground-water flow systems and the depth of circulation can be made from knowledge of the geologic framework of the system, chemistry of the water, flow rate, heat flow, and thermal conductivity of the rocks [33, 34, 35]. Thermal springs discharging from convective hydrothermal flow systems at temperatures greater than 90° Celsius are common in the western United States [36]. They are commonly associated with late Tertiary volcanic and tectonic activity. Convective hydrothermal systems in the tectonically stable eastern United States are of lower temperature (less than 90° Celsius) and are related to deep circulation of ground water [37].

Geochemistry of Ground Water

The chemistry of water at a point in the aquifer is a function of the history of that water in the system. It can be stated in general terms that the dissolved-solids concentration of water increases with the length of the flow path, time of water in the system, and solubility of the minerals in the system. Chemical composition of the water is a function of the sequence of rocks and their chemical constituents with which the water has been in contact and the evolving quality of the water from the time of its entry into the system. The chemical character of the water is thus a product of interrelated chemical and physical phenomona and the geologic framework of the system. Chemical quality can be used as a tool, in conjunction with data on the geologic framework, potentiometric head, and water temperature, in delineating boundaries of flow systems, inferring paths of travel, delineating recharge areas, and estimating maximum reservoir temperatures and age of waters.

Water Budgets

The principal of conservation of mass, or continuity equation, expresses the fact that the inflow of water to a system is equal to the outflow plus or minus the change in storage. In a steady state system, such as a natural, undeveloped system, over a long period of time, the change in storage is negligible and the inflow equals the outflow. This balance is virtually the basis for water budgets. Inflow, or recharge, to systems is largely from precipitation, either by direct infiltration or from the ephemeral flow of streams or ephemeral lakes. Discharge is largely by evaporation or transpiration from the water table, where it is shallow, by springs, or by seeps to lakes and streams. Independent measurements or estimates of the recharge and discharge components provide mutual checks on reliability of the values. Water budgets for flow systems or components of the budget, such as spring flow, evapotranspiration, and infiltration of precipitation, can be readily utilized in mathematical models of the flow system as an integral part of the system hydrodynamics.

CONCLUSIONS

In evaluating geohydrologic factors at the Province level, the factors of paramount importance are the occurrence of potential host rock and delineation of flow system boundaries. Potential host rocks includes salt deposits, tuffs, basalts, crystalline igneous rocks, and shales. Other rocks that may be potential host rocks include unsaturated rocks and alluvial deposits, other evaporites, some metamorphic rocks, and mafic igneous rocks. The waste form can be selected or packaged and engineering design of the repository can be planned to be compatible with the physical and chemical characteristics of a broad range of potential host rocks.

The occurrence of potential host rocks is widespread throughout the conterminous United States. No judgements on the suitability of various host rocks are made in this report. Ultimately, the suitability of the occurrences will depend on the character of the host rock and the favorability of other factors combined to provide multiple natural barriers for the prolonged isolation of radioactive waste from the accessible environment.

The objective of geologic disposal of radioactive wastes is to preclude their reaching the accessible environment until they have decayed to a point at which their concentrations no longer constitute a health hazard. The ground-water flow system is the most likely means by which radionuclides will migrate and eventually reach the accessible environment, if a leak from a repository occurs. Knowledge of the flow systems in which the repository is located will then enable an assessment of the risk, should a leak occur. But, more importantly, knowledge of the geologic framework and flow systems of a Province will enable evaluation of the suitability of many sites that may offer multiple natural barriers to radionuclide transport.

Delineation of flow systems needs to begin in the Province stage of screening, followed by refinement of flow-system delineation and quantitative definition of the hydrologic and geochemical characteristics of the flow system during later stages of screening. The existing data bases on geology and hydrology of the provinces is incomplete, particularly data at depth, where mined repositories might be considered. Initial delineations of flow systems will be imperfect; however, the flow system basis for land unit division provides a framework within which pertinent, available geologic and hydrologic data can be integrated in a flow model. Modeling uses incomplete data on the hydrogeologic framework, potentiometric head, water chemistry, and temperature from a flow system by the use of relating functions, and enables the development of flow system definition from a working hypothesis.

Predictions of the fate of emplaced waste need to take into account hydraulic and chemical characteristics of the rocks and ground water that affect transport of waste by ground water and the rate at which waste comes into solution. Hydraulic and chemical characteristics that need to be considered include: (1) The convective transport by flow, the flow rate in turn being a function of permeability, porosity, and hydraulic gradient; (2) the dispersive and diffusive transport of the contaminant, which may exceed in velocity the average water velocity; (3) the chemical interactions between waste and rocks along the flow path, and the effects of sorption that would delay movement of the contaminant; (4) the length and direction of the flow path, and the porosity, permeability, and potentiometric head gradient along the flow path, factors that determine the rates of flow along flow paths in the system, and the time of flow from a proposed repository to the biosphere; and (5) the relative quantities of fluid in the flow system, which can provide dilution of wastes by ground water.

REFERENCES

1. Interagency Review Group on Nuclear Waste Management: "Report to the President by the Interagency Review Group on Nuclear Waste Management", TID 29442, National Technical Information Service, Springfield, Virginia (1979).

2. Bredehoeft, J. D., England, A. W., Stewart, D. B., Trask, N. J., and Winograd, I. J.: "Geologic Disposal of High-Level Radioactive Wastes, Earth Science Perspectives", U.S. Geological Survey Circular 779, 15 p. (1978).
3. Office of Nuclear Waste Management, U.S. Department of Energy and Geological Survey, U.S. Department of the Interior: "Earth Science Technical Plan for Disposal of Radioactive Waste in a Mined Repository," DOE/TIC 11033 Draft, 68 p. (1980).
4. U.S. Department of Energy and U.S. Geological Survey, Department of the Interior, Subgroup I of Earth Science Technical Plan Working Group: "Plan for Identification and Geological Characterization of Sites for Mined Radioactive Waste Repositories", U.S. Geological Survey Open-File Report 80-686, 73 p. (1980).
5. Bedinger, M. S.: "The Geological Survey Program for Identification of Suitable Geohydrologic Environments for the Disposal of High-Level Radioactive Wastes," Proceedings of the 1980 National Waste Terminal Storage Program Information Meeting, Office of Nuclear Waste Isolation, Columbus, Ohio, Report ONWI-212, pp. 280-281 (1980).
6. U.S. Nuclear Regulatory Commission: "10 CFR Part 60, Disposal of High-Level Radioactive Wastes in Geologic Repositories, Technical Criteria," Draft of Proposed Rule, 61 p. (1981).
7. National Academy of Sciences, National Research Council: "Geological Criteria for Repositories for High-Level Radioactive Wastes", National Academy of Sciences, Washington, D.C., 19 p. (1978).
8. Office of Nuclear Waste Isolation Staff: "NWTS Criteria for the Geologic Disposal of Nuclear Waste, Site Qualification Criteria", ONWI-33(2) Draft, 9 p. (1980).
9. Bayley, R. W., and Muehlberger, W. R., compilers: "Basement Rock Map of the United States", U.S. Geological Survey Map, scale 1:2,500,000, 2 sheets (1968).
10. Smedes, H. W.: "Rationale for Geologic Isolation of High-Level Radioactive Waste, and Assessment of the Suitability of Crystalline Rocks", U.S. Geological Survey Open-File Report 80-1065, 57 p. (1980).
11. Pettijohn, F. J.: "Sedimentary Rocks (2nd ed.)", New York, Harper and Brothers, 718 p. (1957).
12. Yaalon, D. H.: "Mineral Composition of the Average Shale", Clay Minerals Bulletin, v. 5, no. 27, pp. 31-36 (1962).
13. Grim, R. E.: "Clay Mineralogy (2nd ed.)", New York, McGraw-Hill, 596 p. (1968).
14. Merewether, E. A., Sharps, J. A., Gill, J. R., and Cooley, M. E.: "Shale, Mudstone, and Claystone as Potential Host Rocks for Underground Emplacement of Waste", U.S. Geological Survey Open-File Report, 44 p. (1973).
15. Tourtelot, H. A.: "Preliminary Investigation of the Geologic Setting and Chemical Composition of the Pierre Shale Great Plains Region", U.S. Geolgical Survey Professional Paper 390, 74 p. (1962).
16. Smith, G. I., Jones, C. L., Culbertson, W. C., Ericksen, G. E., and Dyni, J. R.: "Evaporites and Brines" in "United States Mineral Resources", U.S. Geological Survey Professional Paper 820, pp. 197-216 (1973).
17. Johnson, K. S., and Gonzales, S.: "Salt Deposits in the United States and Regional Geologic Characteristics Important for Storage of Radioactive Waste", Y/OWI/SUB-7414, Prepared for Office of Waste Isolation, Oak Ridge, Tennessee (1978).
18. King, P. B., and Beikman, H. M.: "The Cenozoic Rocks, A Discussion to Accompany the Geologic Map of the United States", U.S. Geological Survey Professional Paper 904, 82 p. (1978).
19. Dean, W. E., Jr., and Johnson, K. S.: Written Communication, U.S. Geological Survey, Denver, Colorado (1981).
20. Winograd, I. J.: "Radioactive Waste Storage in the Arid Zone", EOS, v. 55, pp. 884-894 (1974).
21. Eakin, T. E., "A Regional Interbasin Ground-Water System in the White River Area Southeastern Nevada", Water Resources Research, v. 2, no. 2, pp. 251-271 (1966).
22. Price, Don, Eakin, T. E., and others: "Water in the Great Basin Region, Idaho, Nevada, Utah, and Wyoming", U.S. Geological Survey Hydrologic Investigations Atlas HA-487 (1974).
23. Mifflin, M. D.: Delineation of Ground-Water Flow Systems in Nevada", Desert Research Institute, University of Nevada System, Hydrology and Water Resources Publication no. 4, 53 p. (1968).

24. Scott, B. R., Smales, T. J., Rush, F. E., and Van Denburgh, A. S.: "Water for Nevada", Nevada Department of Conservation and Natural Resources, Water Planning Report 3, 87 p. (1971).
25. Price, Don: "Summary Appraisal of the Water Resources of the Great Basin", Rocky Mountain Association of Geologists and Utah Geological Association Guidebook, 1979 Basin and Range Symposium, pp. 353-360 (1979).
26. Hanshaw, B. B., and Zen, E-an: "Osmotic Equilibrium and Overthrust Faulting", Geological Society of America Bulletin, v. 76, pp. 1379-1386 (1965).
27. Hanshaw, B. B., and Bredehoeft, J. D.: "On the Maintenance of Anomalous Fluid Pressures II, Source Layer at Depth", Geological Society of America Bulletin, v. 79, pp. 1105-1120 (1968).
28. Wallace, R. H., Jr., Kraemer, T. F., Taylor, R. E., and Wesselman, J. B.: "Assessment of Geopressured-Geothermal Resources in the Northern Gulf of Mexico Basin", U.S. Geological Survey Circular 790, pp. 132-155 (1979).
29. Spencer, C. W.: Personal Communication, U.S. Geological Survey, Denver, Colorado (1981).
30. Berry, F. A. F.: "Hydrodynamics and Geochemistry of the Jurassic and Cretaceous Systems in the San Juan Basin, Northwestern New Mexico and Southwestern Colorado", Stanford University, Stanford, California, Thesis, unpublished (1959).
31. Hanshaw, B. B., and Hill, G. A.: "Geochemistry and Hydrodynamics of the Paradox Basin Region, Utah, Colorado and New Mexico", Chemical Geology, v. 4, pp. 263-294 (1969).
32. Sorey, M. L.: "Numerical Modeling of Liquid Geothermal Systems", U.S. Geological Survey Professional Paper 1044-D, 25 p. (1978).
33. Olmsted, F. H., Glancy, P. A., Harrill, J. R., Rush, F. E., and Van Denburgh, A. S.: "Preliminary Hydrogeologic Appraisal of Selected Hydrothermal Systems in Northern and Central Nevada", U.S. Geological Survey Open-File Report 75-56, 267 p. (1975).
34. Sorey, M. L., Lewis, R. E., and Olmsted, F. H.: "The Hydrothermal System of Long Valley Caldera, California", U.S. Geological Survey Professional Paper 1044-A, 60 p. (1978).
35. Bedinger, M. S., Pearson, F. J., Reed, J. E., Sniegocki, R. T., and Stone, C. G.: "The Waters of Hot Springs National Park, Arkansas, Their Nature and Origin", U.S. Geological Survey Professional Paper 1044-C, 33 p. (1979).
36. Brook, C. A., Mariner, R. H., Mabey, D. R., Swanson, J. R., Guffanti, Marianne, and Muffler, L. J. P.: "Hydrothermal Convection Systems with Reservoir Temperatures $\geq$ 90°C", U.S. Geological Survey Circular 790, pp. 18-85 (1979).
37. Sammel, E. A.: "Occurrence of Low-Temperature Geothermal Waters in the United States", in "Assessment of Geothermal Resources of the United States", U.S. Geological Survey Circular 790, pp. 86-131 (1979).

DISCUSSION

L.J. ANDERSEN, Denmark

In your paper you gave a list of the various criteria of the host rocks. However, as mentioned by Dr. Gera, the indications are only qualitative : sufficient depth, sufficient thickness, etc. and not quantitative. Have you made any attempt to make these criteria more quantitative ?

M.S. BEDINGER, United States

Ultimately, more specific criteria are necessary for evaluation of geohydrologic environments. Criteria are useful as guidelines to preliminary evaluations of suitability, but it is unlikely and unnecessary that a site be ideal with respect to all criteria. The ultimate judgement of suitability will be based on the capability of the system as a whole to isolate waste. Further, criteria are to a certain extent a function of the geohydrologic characteristics of the area under consideration. In the U.S. Geological Survey's screening program, the Province Working Group, will develop the quantitative criteria to be used in screening the Province under study.

A MULTI-PACKER TECHNIQUE FOR INVESTIGATING RESISTANCE TO FLOW THROUGH FRACTURED ROCK AND ILLUSTRATIVE RESULTS

P.J. BOURKE, Chem. Tech. Div., Atomic Energy Research Establishment, Harwell

A. BROMLEY, Geol. Dept., Camborne School of Mines, Cornwall

J. RAE., Theo. Phys. Div., Atomic Energy Research Establishment, Harwell

K. SINCOCK, Geol. Dept., Camborne School of Mines, Cornwall

ABSTRACT

A multi-packer technique was used to locate twelve discrete fractures in the lower half of a 200 m deep drill hole in Cornish granite. The resistances to water flows into these fractures both singly and together were measured. Geological explanations of the results obtained were sought by examination of core from the hole. Analysis of the results and the further data needed and now being sought to determine resistance to flow over long distances through the pattern of interconnected fractures are discussed.

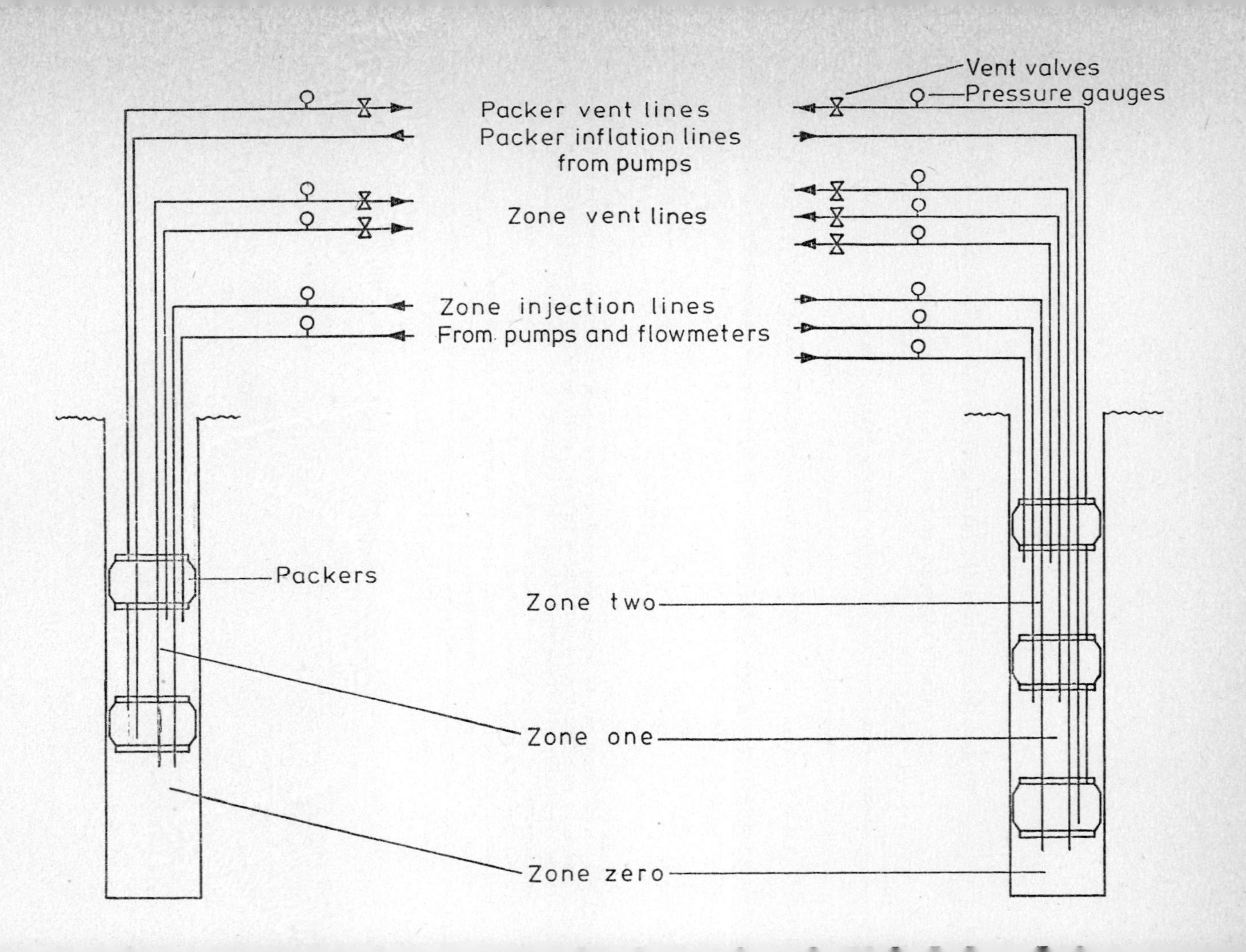

Vent valves
Pressure gauges
Packer vent lines
Packer inflation lines
from pumps
Zone vent lines
Zone injection lines
From pumps and flowmeters
Packers
Zone two
Zone one
Zone zero

1. INTRODUCTION

If radionuclides leak from their containers in a deep depository for radioactive waste in fractured rock such as granite, transport by water movement through fractures is the most likely means by which they could be returned to the surface[1, 2] These fractures are thought to form an extensive, frequently interconnected pattern which, at least in wet climates, is filled with water. There is therefore a need to determine pressure gradients in the water and resistance to flow through the fracture pattern, so that rates and times for flow from the depository to the surface can be predicted. This information, together with knowledge of the retardation of the radionuclide movement behind the water movement by diffusion into stagnant water pores and sorption by the rock, is required for the assessment of the safety of burial of waste.

Pressure gradients causing flow, due both to the slope of the water table and to thermal bouyancy produced by radioactive decay heating, have been considered in previous papers.[3-6]

Resistance to flow was the subject of an earlier Nuclear Energy Agency workshop on lowly permeable rock which was helpful to experimenters planning field measurements for the assessment of potential depository sites. A comprehensive review of permeability data obtained for all purposes was published recently by Brace.[7]

At present there is discussion but no general agreement about whether most of the flow from depositories to the surface will be through a few big fractures or a large number of small ones. If the former is the case, the hydraulic conductivity of each of the individual fractures will be required to predict flow. This information will have to be specifically obtained at each site after the big fractures have been mapped from depository to surface - this may not however be easy. If the latter is the case, more generalised information and a statistical approach will be required, since complete mapping of all small fractures will be impractical.

This paper describes a multi-packer technique which had been previously suggested[8] for investigating the resistance to water flow from a drill hole into both single fractures and groups of fractures. Results obtained in a 200m deep hole in granite are presented and exemplify the use of the technique for studying the hydraulic conductivity of potential depository sites.

A limitation of this technique is that the flows it measures from the hole into the rock may be mainly dependant on the resistance to flow in short lengths of the fractures immediately adjacent to the hole, whereas the required information is the resistance to flow over long distances through the fracture pattern. Both theoretical and further experimental attempts to overcome this limitation are discussed.

2. EQUIPMENT USED

The double and triple packer systems used are sketched in Figure 1. The specially made packers (from Tigra Tierra Inc.,Puyallup, Wa., USA) have hydraulic connections (of 4 MPa, 5mm bore, 1.5mm thick nylon tubing) as shown in the sketches.

The inflation line delivered water through the upper packers directly to the bottom of the lowermost. This ensured that all air could be purged from the packers through the vent line connecting them as shown, by pumping into the inflation line until water issued from the vent valve. The packers were inflated to set them in the hole by closing the vent valve. Their inflation pressure was then determined without uncertainty about air pockets in the lines by adding the pressure measured on the surface before the vent valve to the hydrostatic head in the vent line.

Separate injection and vent lines were connected to each of the zones between and below the packers. All air was again purged by pumping until water issued from the vent lines and tests were started by closing the vent valves. The zone pressures were again determined from the vent pressure gauges and the heads in the lines.

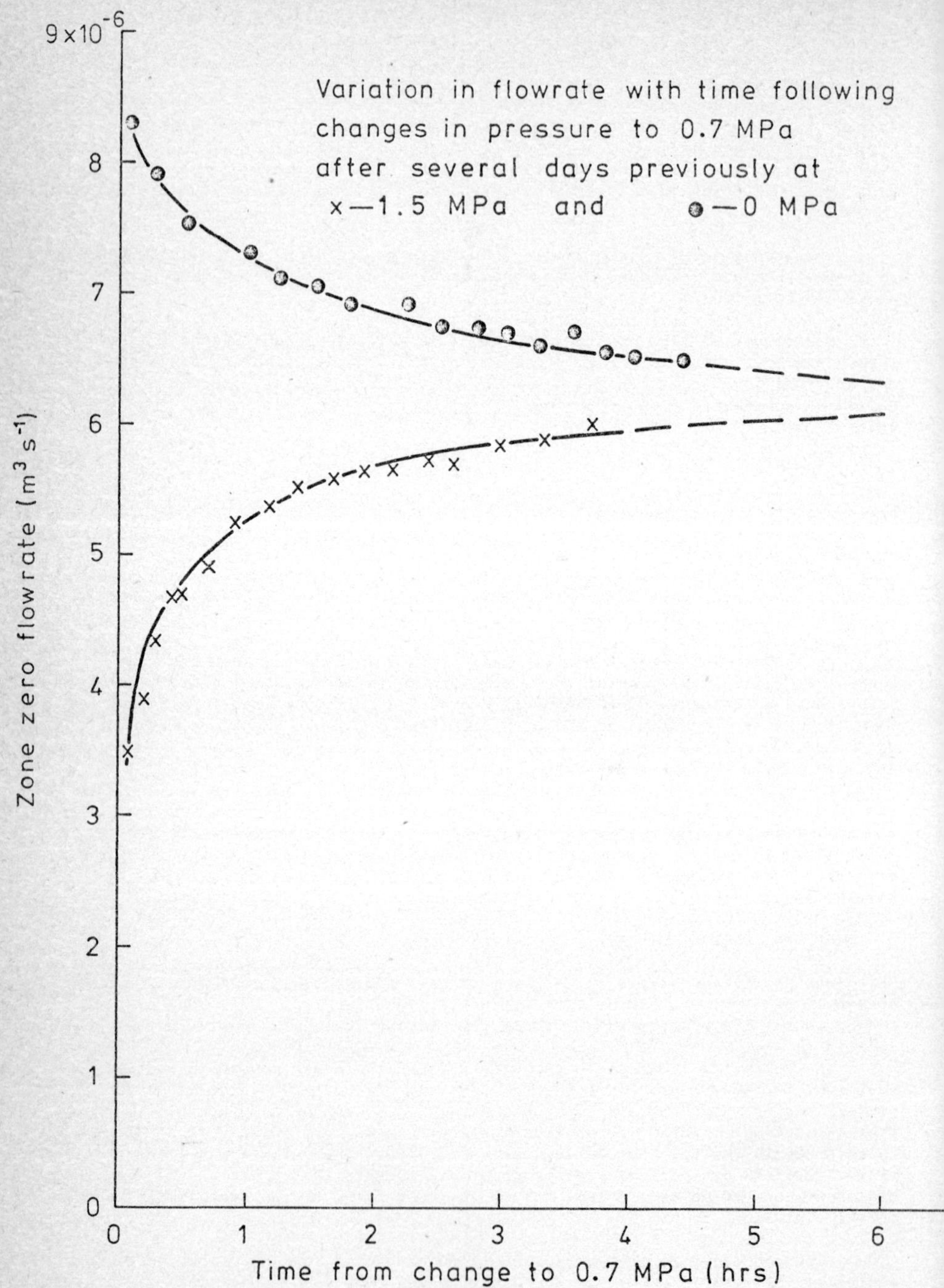

FIG.2 VARIATION IN FLOWRATE WITH TIME AND PREVIOUS PRESSU

The packers could be spaced to any required distances with steel rods. They were then lowered into the hole with steel wire, the length of which was measured by rollers as it entered the hole to determine the depths of the packers.

All pumps used had two connected pistons in separate cylinders. Compressed air moved one piston and the other then pumped water. Water pressures at any value up to 3 MPa could be selected and kept constant indpendently of flow rate up to 3 l/min by choosing an appropriate air pressure.

The rates and totals of water flow into each injection line were separately measured by turbine flow meters (Litre Meter Ltd., Aylesbury, Bucks). These were calibrated after installation and found capable of measuring with moderate accuracy down to 0.1 l/min. Lower flow rates were calculated from the pressure drop along the injection line (i.e. the difference between the injection and vent line pressure gauges on the surface). This calculation was made by interpolating between zero flow and the higher measured flows, using the usual data(9) for pressure drop in pipes.

There were two reasons for choosing these arrangements of packers and hydraulic connections. Firslty, the total flow into all fractures intersecting the length of hole below the packers can be determined without uncertainty about leakage past the lower of the double packers. This is ensured by maintaining equal pressures below (zone zero) and between (zone One) the packers. There can then be no leakage past the lower packer and any leakage past the upper will not appear in the metered flow to zone zero. Secondly, the separate flow into any individual fracture intersecting zone one can similarly be determined without uncertainty about leakage using the triple packers.

3. EXPERIMENTAL PROCEDURE

Experiments were made as follows. After purging the lines, setting the packers with 3 MPa inflation pressure, and closing the zone vent valves; the water pressures in the zones was maintained equal and constant and the flow rates were monitored until they too become constant. In practice this monitoring could take an inconveniently long time; up to many hours to approach closely to a steady value. See Figure 2 for typical variations in flow rates with time at constant zone pressure following periods of days during which higher and lower pressures had been maintained in the zone. Clearly the approach to steady flow depends on the previously established pressure field through the fractures.

This lead to the adoption of the following procedure to shorten the time to make experiments. Immediately prior to the monitoring for the steady flowrate at any zone pressure, a higher or lower pressure, depending on whether the zone had previously had a lower or higher pressure respectively, was held for a short time. The time taken to obtain steady flowrate was thereby appreciably shortened.

For each depth of the packer system, the steady flowrate was measured at each of several different zone pressures. The highest zone pressure was however always limited to be about 1 MPa below the packer inflation pressure to avoid any leakage past the packers or interference between zones.

4. RESULTS OBTAINED

The packer systems described above were used in a 200m deep, 100mm diameter hole in Cornish granite. The packers with a 3m long zone one were first lowered to make measurements at the bottom of the hole and then raised in 3m steps for repeat measurements up the lower half of the hole. This ensured that each fracture appeared first in zone one and later in zone zero.

The results presented here were obtained during a four month period while the equipment and experimental procedure were being developed and it should therefore be noted that they are only preliminary results. To check their reproductibility, a careful repeat series of measurements is now being made in what is hoped will be a shorter time.

At every depth, steady flow rates were quite consistently found to be linearly related to zone pressure. See Figure 3 for typical variations in zone

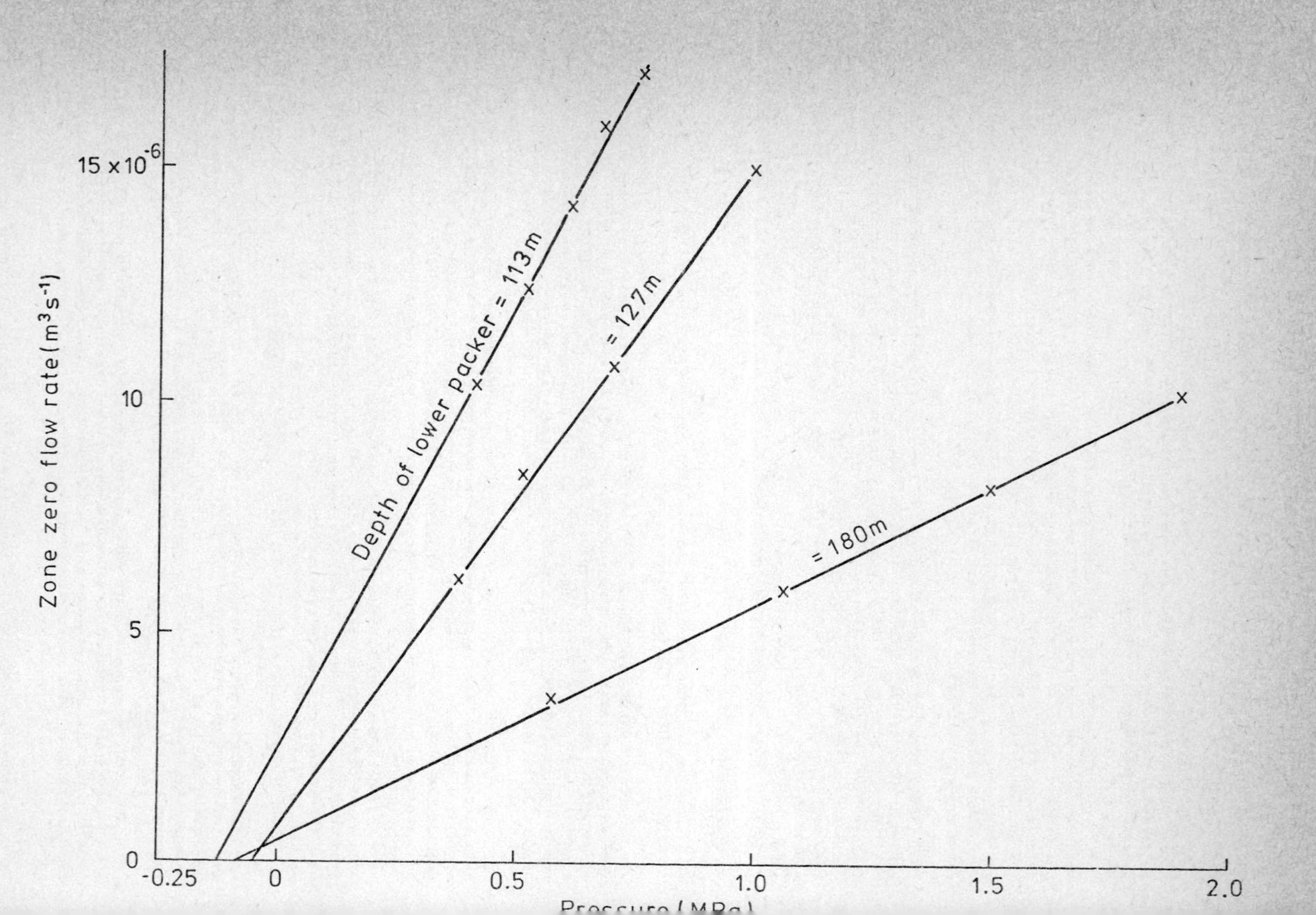
Zone zero flow rate (m3s-1)
15 x 10-6
10
5
0
-0.25
0
0.5
1.0
1.5
2.0
Depth of lower packer = 113m
= 127m
= 180m

zero flow rates plotted against the zone pressure indicated by the pressure gauge in the vent line at the surface. Extrapolation of these straight line plots to zero flow shows the corresponding pressures to be negative on this scale. This can be explained by the natural undisturbed water level's being below the surface. In fact the average and standard deviation of such presures deduced for zero flow at all depths tested were about -0.1 MPa and ± 0.05 MPa. This is closely consistent with the 10m average depth below the surface of the water table which would give a surface pressure of -0.11 MPa.

These observations give support to two important assumptions made in analysing the results. Firstly, that the fracture pattern is connected to the surface and filled with water from the water table downwards. Secondly, that the flow into the fractures at all depths tested is laminar and proportional to pressure in excess of the head produced by the water table.

For every depth the average values of the steady flowrates ($m^3 s^{-1}$) from both zones zero and one were divided by the excess pressures (MPa) causing flow and are plotted against depth below the surface in Figure 4. For zone zero the depth was taken to be that of the lowermost packer and for zone one it was taken to be that of the middle of the zone.

The flowrates per unit pressure from zone one show that there are twelve positions below 80m depth in the hole at which appreciable flows enter the rock. Between these there are lengths of no flow measurable with the $\simeq 10^{-7} m^3 s^{-1}$ MPa^{-1} sensitivity of the equipment. These observations are consistent with the increases in flowrates per unit pressure found in zone zero as the packers were raised. Together both sets of measurements show that most of the flow enters the rock in several widely separated, short zones of locally much higher hydraulic conductivity than the rest of the rock. These zones must presumably contain either discrete fractures or thin layers of highly permeable rock.

5. GEOLOGICAL COMMENT ON RESULTS

Any geological explanation or recognition of these high flow regions is valuable in analysing the hydraulic results. Regional geology and observations from drill core from the hole therefore are reviewed below.

Geological Setting of Site

The experimental site near Camborne in Cornwall is located near to the north-western margin of the Carnmenellis granite. This granite is one of the six major masses which, together with several minor outcorps, comprise the upper, exposed portions of the Cornubian batholith. This batholith was emplaced into folded and regionally metamorphosed Devonian and carboniferous rocks, ca. 285 m.y. ago, after the main deformation phase of the Hercynian orogeny.

There is strong geophysical and geological evidence to suggest that the exposed plutons rise from a continuous ridge of granite which extends from Dartmoor to beyond the Scilly Isles. The model proposed by Bott(10) is of granitic material with a seismic velocity of 5.85 km sec^{-1} extending downwards to a depth of about 12km. This grades into lower crust, of presumed intermediate composition, having a mean velocity of 6.5 km sec^{-1} and which is in turn, underlain by mantle rocks at a depth of about 27 km.

The bulk of the rocks which make up the exposed parts of the batholith are either coarse or medium grained biotitie-muscovite granites with or without megacrysts of potash feldspar. Lithionite graphites, both prophyritic and non-porphyritic, make up parts of the Godolphin-Tregonning and the St. Austell masses. Fluorite granites consititute parts of the St. Austell intrusion and all the larger bodies carry masses of mafic-deficient leuco-granite, small dykes and veins of aplite and simple pegmatites. The associated dyke phase, together with the small cupolas of St. Michael's Mount and Cligga Head comprises both quartz-feldspar porphyries ('elvans') and fine grained, flow banded rhyolites. Post consolidation alteration, which has affected all of the exposed masses to a greater or lesser extent, includes tourmalinisation, greisening, choritisation, haematisation and both widespread and localised kaolinisation. The extensive hydrothermal vein

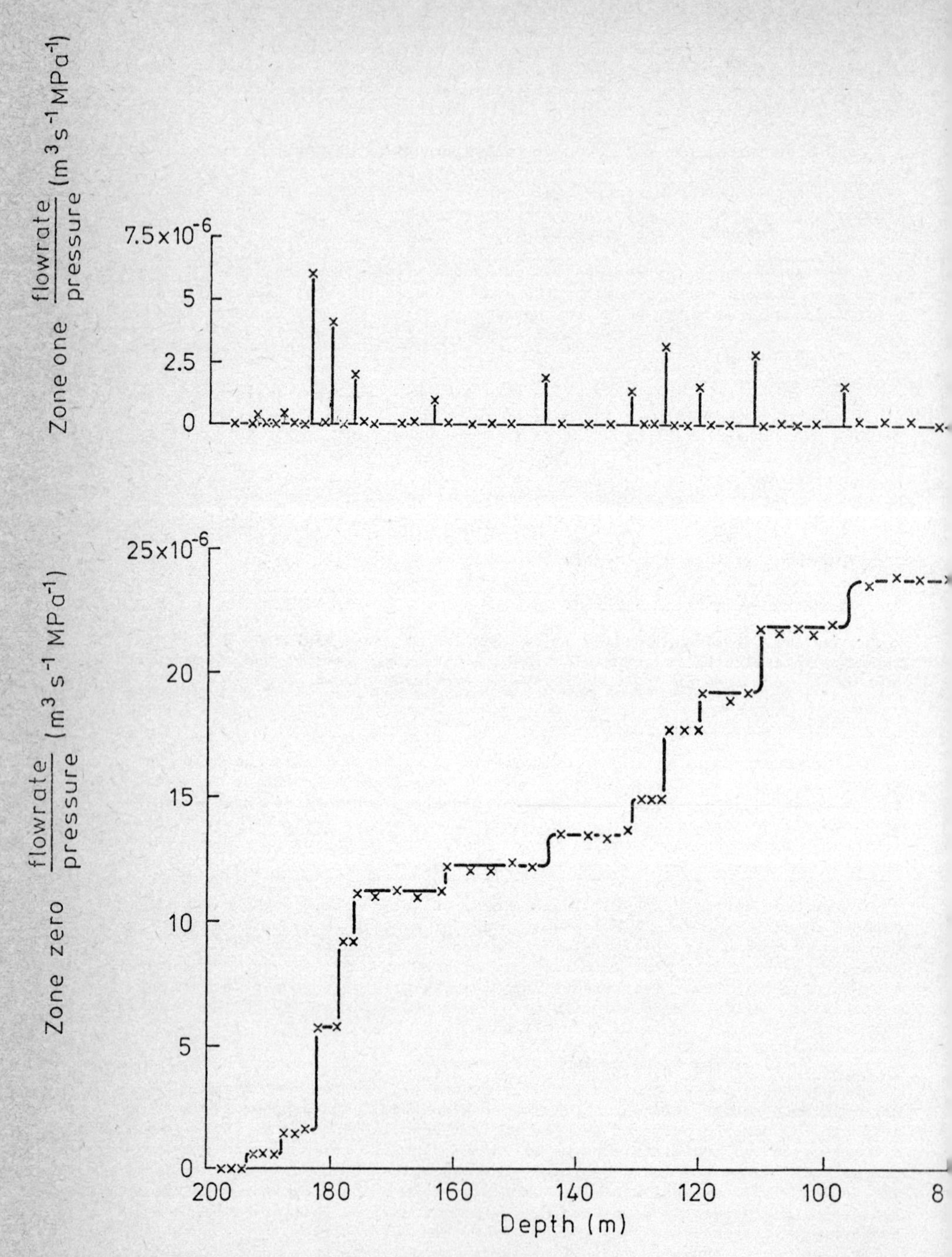

FIG. 4 FLOWRATES PER UNIT PRESSURE INTO ZONES ONE AND ZE

mineralisation, which includes tin, copper, tungsten and a wide range of other metals of lesser importance, is spatially and probably genetically associated with the granites. The most intensely mineralised area in S.W. England, between the towns of Camborne and Redruth lies 2-3 km north west of the experimental site.

The site is located about 1 km inside the north western margin of the Carnmeneliis mass. The principal rock type exposed at the surface and encountered in boreholes is a coarse grained (5 mm) feldspar porphyritic, biotite-muscovite granite. The granite invariably shows incipient hydrothermal alteration which has resulted in partial to complete chloritisation of biotite and incipient kaolinisation of feldspars. Albitites, never recorded at the surface, have been encountered at two levels in boreholes; granite porphyry dykes ('elvans'), varying in thickness between a few tens of millimetres and several metres are known from surface outcrops and in the boreholes.

In the quarry at the site and in adjacent, shallow underground workings two major sets of steeply dipping fractures with strike maxima at 103° and 194°, are recognised. They have spacings of the order of one metre and may have clean, closely matching walls, or may be lined by variable thicknesses of secondary minerals such as chalcedony and haematite and bordered by zones of kaolinised or haematised granite. Sub-horizontal fractures have spacings of less than one metre immediately below the present land surface but decrease in frequencey downwards. Below 100m depth their spacing is of the order of 5-10 metres. Rare mineralised veins with sulphides and cassiterite and most 'elvan' dykes strike at approximately 050°, parallel with the regional trend, while fractures with moderate dips (45-70°) have very variable strike directions but are mush less frequent. Like the steeply dipping fractures, they may have closely matching walls or be more or less open, lined with soft secondary minerals and bordered by zones of altered granite.

Observations from Examination of Core from Hole

Detailed logging has been carried out on diamond drill core from the hole in which the hydraulic results were obtained in an attempt to correlate zones of high flow with geological features which might be responsible for this.

Fractures found in the core were classified according to the following criteria:

- dip angle from horizontal
- apperture; defined as
  - narrow where the core locked tightly together
  - moderate where the core did not lock but feldspar crystals matched across the fracture
  - wide where the feldspar crystals did not match across the fracture
- lithology; i.e. presence of minerals filling the fracture and alteration of granite adjacent to fractures.
- other potential aquiferous features; e.g. zones of altered granite, 'elvan' dykes, cavernous pegmatite veins etc.

Seven major zones of fractures which may cause high conductivity were found in the core, as shown in the table here.

Depth (m.)	Potentially Aquiferous Features
120.5 - 124	Steeply dipping 'elvan' dyke with porous alteration zones and many steeply dipping fractures from 121 to 123.5 m.
150 - 154.5	Several steeply dipping fractures with highly mismatching walls and bordered by kaolinised granite. Thin fractured 'elvan' dyke at 156 m.

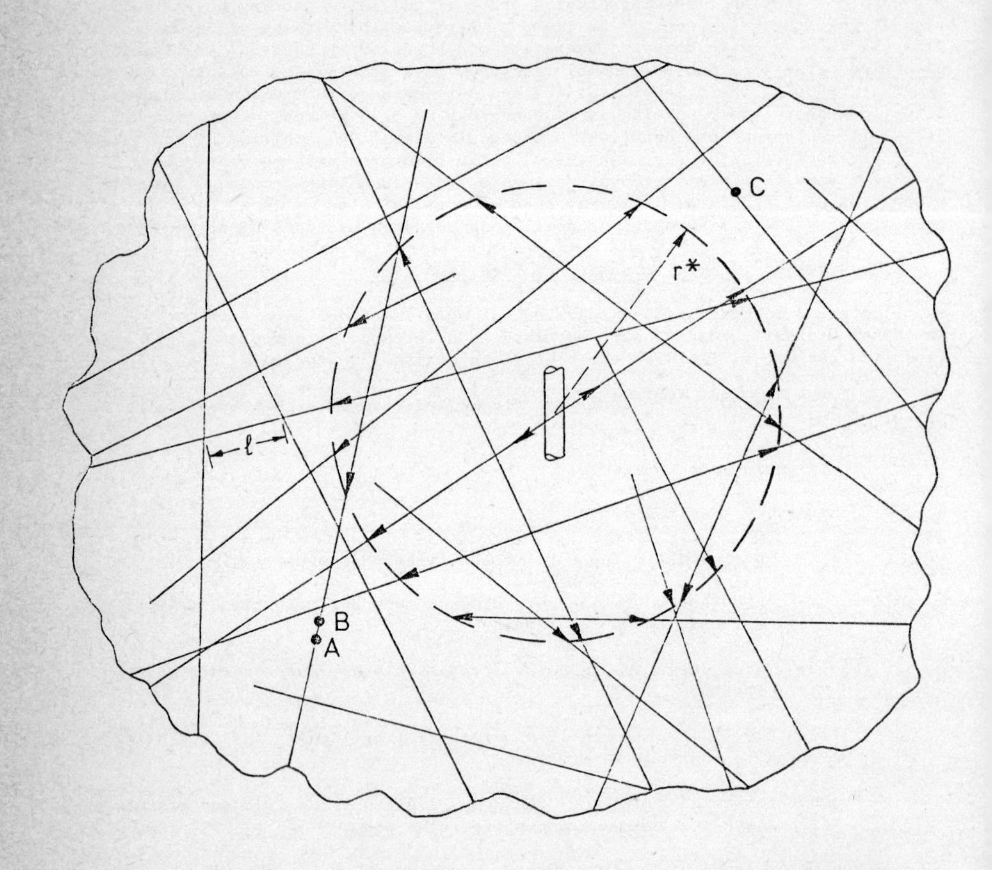

FIG. 5. A POSSIBLE SECTION THROUGH FRACTURE PATTERN.

Depth (m.)	Potentially Aquiferous Features
160 - 164	Three flat lying fractures with mis-matching walls and bordered by kaolinised granite at 160, 160.5, and 164 m.
175 - 178	Mis-matching, steeply dipping fracture with irregular chalcedony filling at 177 m. Two steeply dipping, mis-matching fractures at 177.5 m.
182 - 183.5	Non-matching, flat lying fracture at 182 m. Steeply dipping kaolinised fracture. Non-matching at 183 m.
187.5 - 189	Three moderaltey dipping, non-matching kaolinised fractures from 187.5 to 188 m. Steep clay-filled fracture at 189 m.
194 - 194.5	Largely intact core. Very steep, non-matching clay-filled fracture from 194 to 194.5 m.

Six of these zones match the zones of high flow. Five of them exhibit steeply (more than 80°) dipping, potentially aquiferous features which include two elvan dykes, two wide apperture fractures with kaolinised borders and one open fracture partly lines with chalcedony. One zone had only moderate dipping fractures and another has only flat lying ones.

No significant statistical correlation of zones of high flow can be made with the variation in frequency of all fractures found along the core. It is however tentatively concluded that several different geological features (kaolinised fractures, elvans etc) may have been the causes of flow into the rock and that both the appertures and dips of these features are important. Steep dip may have produced high flow for two reasons. Firstly, near vertical fractures tended to have wider appertures than those closer to the horizontal. Secondly, the longer perimeters of elliptical intersections of near vertical fractures with the vertical hole allowed flow into the fractures with lesser resistance than the shorter perimeters of the circular intersections of horizontal fractures (as quantified by equation (1) below).

6. ANALYSIS OF RESULTS

The question of how to represent resistance to flow through fractures is raised by considering the geometry of their pattern. If it is assumed that the fractures are planes which are randomly oriented in space, a section through the pattern could be as shown in Figure 5. If one considers any pair of points, such as A and B, separated in one fracture by a length, small compared with the distances, typically 1 in Figure 5, separating intersections between fractures, then the resistance to flow between these points will be mainly dependant on the local apperture and hydraulic conductivity of this one fracture. If alternatively one considers another pair of points such as B and C separated by a length large compared with the average separation between fracture intersections, the the resistance to flow between these points will depend on the hydraulic conductivities of the many paths joining them and may realistically be represented by the effective permeability of the fracture pattern treated as a permeable medium.

Flows into Single Fractures in Zone One

This consideration immediatley highlights a problem of choosing between two extreme cases in analysing the present measurements of flows from zone one into single fractures. If in the first case, the distance from the hole along the fracture to its intersection with another is so large that most of the pressure drop causing flow occurs along a shorter distance in that fracture, then the measured flow will be mainly dependent on its individual conductivity. The flows into different fractures should therefore in this case exhibit the same variation as that of conductivities. If in the second case, the average separation between intersections is so small and the average conductivity is so large that the pressure drop extends over many fractures, then the measured flow will depend mainly on the effective permeability of the pattern. The flows into the pattern from adjacent fractures in the hole should therefore in this case not vary widely

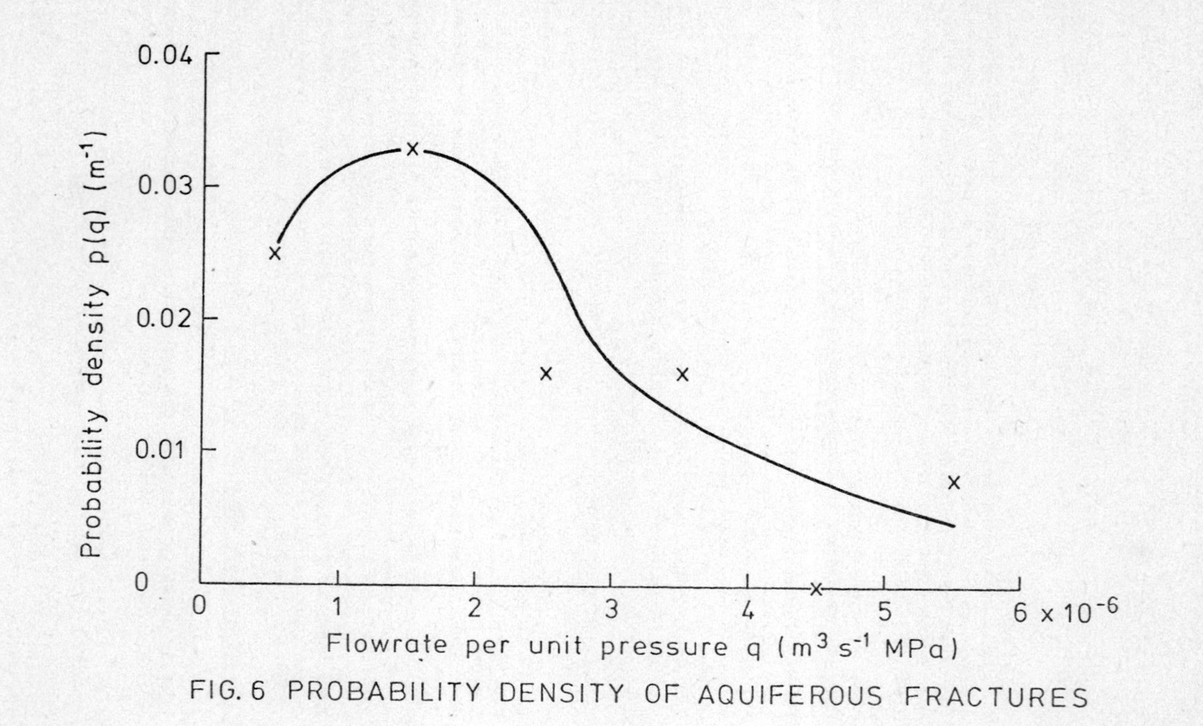

FIG. 6 PROBABILITY DENSITY OF AQUIFEROUS FRACTURES

because they are all mainly determined by the same permeability.

The flows into the twelve fractures found in the hole did in fact vary by more than a factor of ten and this therefore strongly suggests that the first case above is applicable.

This argument leads to the following suggestion for interpreting the results. The flow from the hole will firstly be two dimensional in a single fracture. The steady state equation for this flow rate; Q from a cylindircal hole of radius; R intersected at an angle θ by a planar fracture of apperture; d is given by

$$\frac{Q}{p(R) - p(r)} = \frac{\pi d^3}{6\mu}\left[\ln\frac{r}{R} + \ln\frac{2\cos\theta}{1+\cos\theta}\right]^{-1} \qquad \text{..........(1)}$$

where p = pressure in excess of the hydrostatic head
r = distance from centre of hole

(see appendix for derivation).

As the flow encounters fracture intersections it will change to three dimensional in the fracture pattern and may then be treated as sp herically outward in an extensive permeable medium, as indicated by the arrows in Figure 5. The steady stae equation will then

$$\frac{Q}{p(r)} = \frac{4\pi k}{\mu} \cdot r \qquad \text{..........(2)}$$

where k = permeability.

The change from two to three dimensional flow will occur over a range of radial distances from the hole. If however it is assumed to occur at some representative distance; r* (Figure 5) then as an approximation equations (1) and (2) could in principle be solved simultaneously for the apperture; d and permeability; k, with common values of r* and p(r*).

r* must presumably be a few times greater than the average separation between fracture intersections. If the fracture planes are in fact randomly orientated, their average 10 m separation in the hole suggest that the average intersection separstion will also be about 10 m and r* may therefore be estimated to be some tens of metres.

p(r*) cannot unfortuantely be estimated. Further it may be too small to measure easily since most of the pressure drop causing flow is thought above to occur over smaller distances than r*. It cannot however be zero - otherwise equation (2) would give an infinite value for the permeability k - and it must therefore have a small finite value.

It may be possible to overcome this difficulty if p(r) can be measured experimentally at some intermediate values of r such that $R<r<r^*$, in which case p(r*) may be determined by extrapolation using equation (1). These measurements are being attempted in the continuing programme discussed below, with larger radius holes to give larger values of p(r).

Total Flows into All Fractions in Zone Zero

Attempting to analyse the measurements of total flows into all fractures in zone zero raises the question of how representative the range of the conductivities of these fractures is of all fractures. To answer this question the probability density of the flowrate per unit pressure, q into each fracture was calculated and is plotted as P(q), theprobability per metre length of hole of occurrence of a fracture of q $\pm$ 0.5 X 10^{-6} agianst q ($m^3s^{-1}MPa^{-1}$) in Figure 6.

The scatter of the points clearly shows that the number of data are inadequate to define theprobability variation accurately. Nevertheless the fact that the curve appears to be falling rapidly at each extreme suggests that a representative range of conductivities may have been encompassed.

If this is so, there is some justification for treating the total measured flows as if they were into a permeable medium. In this case the total flow rate; Q_t from a long length; L of hole into a fracture pattern of effective permeability; k is given by; neglecting axial flow end effects and the mean value of the second logrithmic term in equation (1)

$$\frac{Q_t}{p(R) - p(r)} = \frac{2\pi Lk}{\mu} \left[\ln \frac{r}{R} \right]^{-1} \qquad \text{..............(3)}$$

where $R < r << L$

In attempting to use this equation, the need for other pairs of values for r and p(r) is again encountered. On the foregoing argument

$$p(r') << p(R)$$

where r' is the average separation between fracture $\simeq$ 10 m.

Taking r' to be 10 m and neglecting p(r') in comparison with p(R), a minimum possible value of k of $\simeq 10^{-16} m^2$ was found by varying k in equation (3) to obtain the best fit with the experimental data in Figure 3. This should, however, be considered to be only an approximate estimation and more accurate determination of permeability again requires measurement of pressures through the fracture pattern.

7. FUTURE WORK

In the continuing experimental programme, other holes have now been drilled adjacent to that in which the present results were obtained. Radioactive tracers are being used to determine flow paths between holes. Multi-packers will be used in an attempt to measure pressure drops along the fractures, as shown necessary above for rigorous analysis to determine resistance to flow through the fracture pattern. The experimental data will also be increased for better statistical analysis by tests in deeper holes.

8. CONCLUSIONS

A multi-packer technique has been developed for studying the resistance to flow through fracture patterns in rock.

The results obtained from 100 to 200 m deep in Cornish granite have given information about the variation in hydraulic conductivities of single fractures and groups of fractures and indicate that the average separation between the fractures through which most of the flow occurs is about 10 m.

Several geological features, most notably some types of fractures and their apperture and dip angle account for some of the hydraulic results. The fracture pattern has been approximately estimated to have a minimum possible value of $\simeq 10^{-16} m^{-2}$.

Analysis of the results shows that rigorous determination of both the conductivities of individual fractures and the permeability of the pattern require further experimental data, particularly the measurement of pressure drop along fractures which is now being attempted.

9. ACKNOWLEDGEMENTS

The authors wish to thank Mr B Watkins and Mrs D Pascoe for all the experimental work involved often in inclement weather. This research is part of the contribution of the United Kingdom Atomic Energy Authority to the programmes of research of the HMG Department of Environment and the Commission for European Communities into the safety of burial of radioactive wastes.

10. REFERENCES

1. R. Heremans, E. Barbreau, P.J. Bourke and H. Gies, "1980 CEC Radioactive Waste Management Conf. Luxembourg."

2. P.A. Witherspoon, N.G.W. Cook and J.E. Gale, : "1981, Science (in press)."

3. J.L. Ratigan, A. Burgess, E.L. Skiba, and R. Charlwood, : "1977, Groundwater movements around a repository : Repository domain groundwater flow analyses, KBS TR54:05."

4. "Office of Waste Isolation, 1978, Technical Support for GEIS: Radioactive Waste Isolation in Geologic Formations: Groundwater Movement and Nuclide Transport, Y/OWI/IM-36/21."

5. P.J. Bourke and D.P. Hodgkinson, "1979a, NEA/IAEA, Low Permeability Symposium, Paris."

6. P.J. Bourke and P.C. Robinson, "1981, Int. Jrnl. of Radioactive Waste Management (in press) and AERE R 9951."

7. P.J. Bourke, J.E. Gale, D.P. Hodgkinson and P.A. Witherspoon, : "1979b, NEA/IAEA, Low Perbeability Symposium, Paris."

8. W.F. Brace, : "Int J. Rock Min. Sci. Vol. 17 (1980) p 241."

9. J.H. Perry, : "Chemical Engineers Handbook, McGraw Hill."

10. M.P.H. Bott, A.P. Holden, R.E. Long and A.L. Lucas, : 1970 "Crustal structure beneath the granites of south-west England. In Mechansim of igneous instrusion, 93-102. Liverpool".

APPENDIX

FLOW INTO A SINGLE PLANAR FRACTURE FROM ITS OBLIQUE INTERSECTION WITH A PRESSURISED HOLE

We consider a planar fracture of width d intersected obliquiely, at angle θ by a circulat borehole of radius R. The region of intersection is supposed to be filled with water at a maintained pressure P_1 which flows away slowly but steadily into the fracture. Apart from small edge effects, which we will ignore, the region of intersection has the form of an elliptic cylinder

$$\frac{\cos^2\theta}{R^2} x^2 + \frac{y^2}{R^2} = 1, \ -\frac{d}{2} < z < \frac{d}{2}$$

The equations which describe the water flow are the steady Stokes equations[1]

$$\nabla . \underline{u} = 0$$

$$\frac{1}{\mu} \nabla p = \nabla^2 \underline{u}$$

where $\underline{u}$ is the flow velocity, p the pressure and μ the dynamic viscosity. The boundary conditions still have to be set. On the wall of the elliptic cylinder we take $p = p_1$. For a second boundary condition we would like to fix an "ambient" pressure p_o at some suitable reference distance, but some care is needed here as for $\theta \to \frac{\pi}{2}$ the major axis of the ellipse becomes arbitrarily long. For a borehole of length L the simple treatment here only makes sense for angles $\theta < \tan^{-1} \frac{L}{2R}$ and we restrict ourselves to considering these. We then want to choose a reference distance much larger than L. For conveneince we fix the ambient pressure p_o on the elliptical cylinder

$$\frac{\cos^2\theta_o}{r^2} x^2 + \frac{y^2}{r^2} = 1 \qquad -\frac{d}{2} < z < \frac{d}{2}$$

where $\theta_o = \tan^{-1} (\frac{R}{r} \tan \theta)$. This cylinder is confocal with the previous one but for large values of r (compared to L) is indistinguishable from a right circular cylinder of radius R_o.

Now we have a flow region bounded by confocal elliptic cylinders it is convenient to solve the equations in terms of ellipitic cylinder co-ordinates ξ, η, z related to the original cartesian set by[1]

$$x = c \cosh \xi \cos \eta \qquad y = c \sinh \xi \sin \eta \qquad z = z$$

where the "radial" co-ordinate ξ varies $(0, \infty)$ and the "angular" co-ordinate η $(0, 2\pi)$.

To corespond to the above ellipses we need to take $c = r \tan \theta$ and then the inner ellipse is $\xi = \xi_1$ with $\tanh \xi_1 = \cos \theta$ and the outer ellipse $\xi = \xi_o$ with $\tan \xi_o = \cos \theta_o$.

As we have creeping flow the pressure satisfies Laplace's equation and since it is only a function of the "radial" co-ordinate ξ this reduces to[1]

$$\frac{{}^{2}p}{\xi^{2}} = 0.$$

It follows at once that the pressure gradient $\frac{p}{\xi}$ is the constant $\frac{p_1 - p_o}{\xi_1 - \xi_o}$.

We next look for solutions of the flow equations in which only the ξ-component of velocity, we call it u, is non zero. In the new co-ordinates the equation $\nabla.\underline{u} = 0$ then gives(1)

$$\frac{}{\xi}(\sinh^2\xi + \sin^2\eta)^{\frac{1}{2}}u = 0$$

with hindsight we pick the solutions

$$u = \frac{f(z)}{(\sinh^2\xi + \sin^2\eta)^{\frac{1}{2}}}$$

where f(z), a function of z alone, is to be determined from the other flow equation. The expression $\nabla^2\underline{u}$ can be evaluated in elliptic cylinder co-ordinates by writing it(1) as $\nabla^2\underline{u} = \nabla(\bar{\nabla}.\underline{u}) - \nabla x\,(\nabla x\underline{u}) = -\nabla x(\nabla x\underline{u})$. The manipulations are cumbersome but it turns out that the η and z components vanish and the ξ component gives rise to the simple equation

$$\frac{1}{\mu}\frac{p}{\xi} = \frac{d^2 f}{dz^2}$$

As the velocity u must vanish on the walls of the fracture at $z = \pm\frac{d}{2}$ the final solution for the velocity is

$$u(\xi,\eta,z) = \frac{1}{c(\sinh^2\xi + \sin^2\eta)^{\frac{1}{2}}}\;\frac{1}{2\mu}\;\frac{p}{\xi}\;(z^2 - \frac{d^2}{4}$$

To find the total volume flow rate Q this equation has to be integrated round the wall of the elliptic cylinder with respect to the area element $c(\sinh^2\xi_1 + \sin^2\eta)^{\frac{1}{2}}$ $d\eta\, dz$. The result is

$$\frac{Q}{p_1 - p_o} = \frac{\pi d^3}{6\mu\,(\xi_o - \xi_1)}$$

This result can be made more transparent by using the relations between the ξ's and θ's and the smallness of θ_o to write

$$\xi_o - \xi_1 = \tfrac{1}{2}\,\ell n\;\frac{1 + \cos\theta_o}{1 - \cos\theta_o}\;\frac{1 - \cos\theta}{1 + \cos\theta} \simeq \ell n\;(\frac{2R_o}{r}\;\frac{\cos\theta}{1 + \cos\theta})$$

so

$$\frac{Q}{p_1 - p_o}\qquad\frac{\pi d^3}{8\mu}\;\{\ell\eta\;\frac{r}{R} + \ell\eta\;\frac{2\cos\theta}{1 + \cos\theta}\}^{-1}$$

For $\theta \to 0$ the second $\ell\eta$ term drops out and we recover the result for a circular cylinder. As θ gets bigger, towards $\tan^{-1}\frac{L}{R}$, the second $\ell\eta$ term is large and negative and Q is increased by the effect of the angle.

It is also easy to check that when $\theta \to 0$ the expression for the velocity u reduces to the circular cylinder value

$$u(p,z) = \frac{(z^2 - d^2_4)}{2\mu p}\;\frac{p_1 - p_o}{\ell\eta\,\frac{r}{R}}$$

REFERENCE

HAPPEL J. and BRENNER H. Low Reynolds Number Hydrodynamics, (Noordhoff, Leyden) 1973.

DISCUSSION

L.T. ANDERSEN, Denmark

In Denmark we use a four packer device for taking water samples from screened water wells. We pump from the upper and lover zone and collect the sample from the medium zone. In that way we can be sure that no leakage takes place around the packer and that the water sample is from the interval we suppose it is.

SESSION IV

Chairman - Président

M.S. BEDINGER

(United States)

SEANCE IV

HYDRAULIC TESTING OF LOW PERMEABLE CHALK FORMATION, MORS, DENMARK

Gosk, E.
The Geological Survey of Denmark (DGU)
Copenhagen, Denmark

ABSTRACT

The testing of low-permeable formations (k = 0.01-1.0 millidarcies) gives rise to a number of specific problems with regard to the choice of the measuring equipment, the design of the technical arrangement and the evaluation of the results. Five different testing methods used during the hydrological investigation of the chalk formation of the Mors salt-dome area are described:

1) constant-rate pumping tests
2) step drawdown tests
3) step injection tests
4) conventional slug tests, and
5) pressurized slug tests.

Recommendations as to how to choose the testing method, how to design the technical arrangement and how to control the results obtained are given.

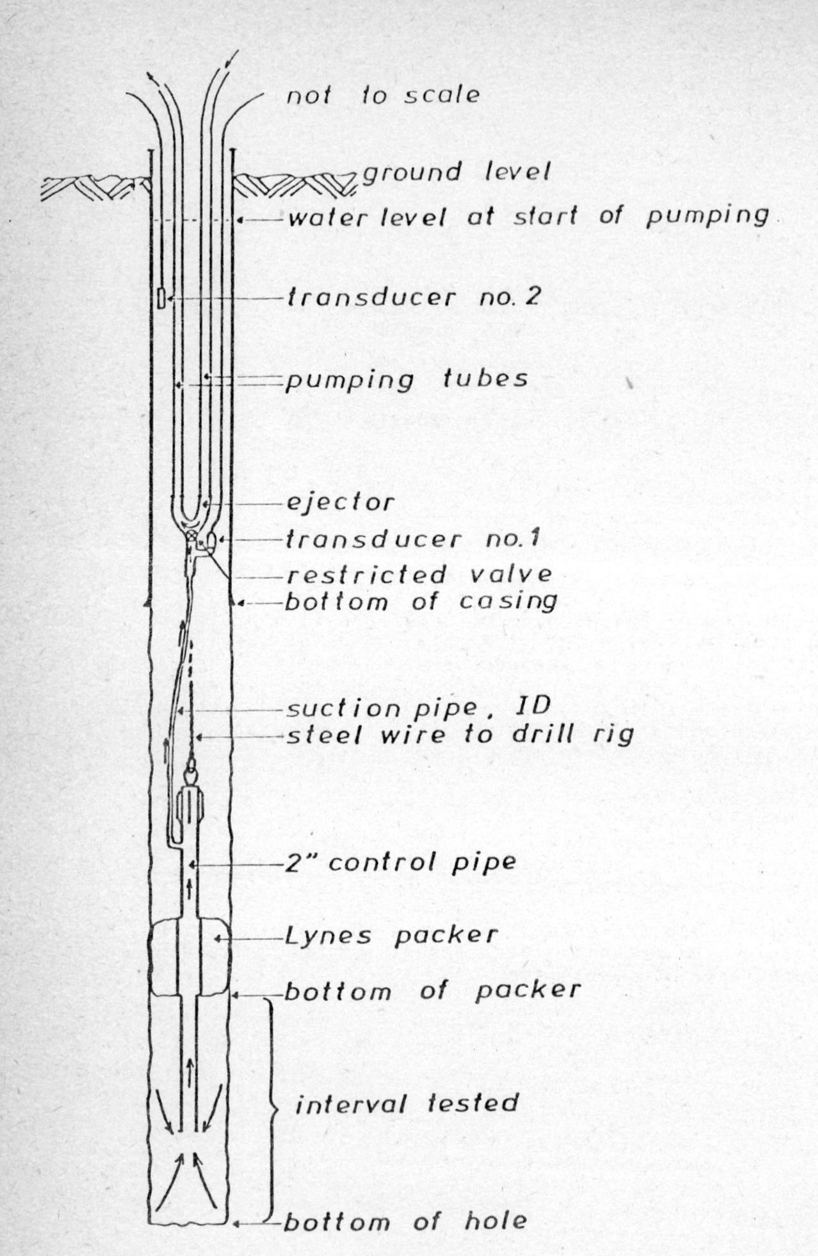

Fig. 1: Sketch of the technical arrangement, bottom hole test.

1. INTRODUCTION

The use of salt domes as repository sites for the final disposal of radioactive waste is considered in Denmark.

An investigation programme designed for the Mors salt dome (North-Jutland, Denmark) consisted of the drilling of two deep salt wells (down to 3500 m b.s.) and four hydrogeological wells, abbreviated E1S, E2S, E3S and E4S, drilled to the depth of about 560 m. The results of the hydrogeological tests are described in DGU report /1/.

This paper presents some of the testing methods used in the hydrogeological wells. Three parallel papers, /2/, /3/, and /4/, deal with other aspects of the investigation. Only a few examples showing typical response are given, and the feasibility and the reliability of the different testing methods are discussed. Recommendations with regard to the choice of the testing method and the design of the technical arrangement are included.

2. TESTING METHODS

2.1 Constant-rate pumping tests

Two types of constant-rate pumping tests have been used: bottom-hole tests (BHT), where the tested interval was defined by the bottom of the packer and the actual bottom of the hole; straddle-packer tests (SPT), where two packers were used to confine the tested interval. The technical arrangement used for BHT is shown in Fig. 1. The technical arrangement used for SPT was similar but for two differences: an extra packer and 50-70 m of 2" perforated pipe were added. SPTs were performed after the total depth was reached, which made it possible to test arbitrarily chosen intervals.

2.1.1 Theoretical background

For idealized conditions the drawdown in the well may be described /5/ by eq. 1:

$$H = 2.3Q/(4\pi T)\log(2.25Tt/(r_w^2 S)) \qquad (1)$$

where:
- H = drawdown in the well (m)
- Q = pumping rate (m^3/sec.)
- T = transmissivity (= permeability x thickness of the pumped interval) (m^2/sec.)
- t = pumping time (sec.)
- r_w = radius of the well (m)
- S = storage coefficient (dimensionless)

Equation 1 is an approximation to the exact solution but, for the occurring parameter values, it may be used after 1-10 minutes of pumping.

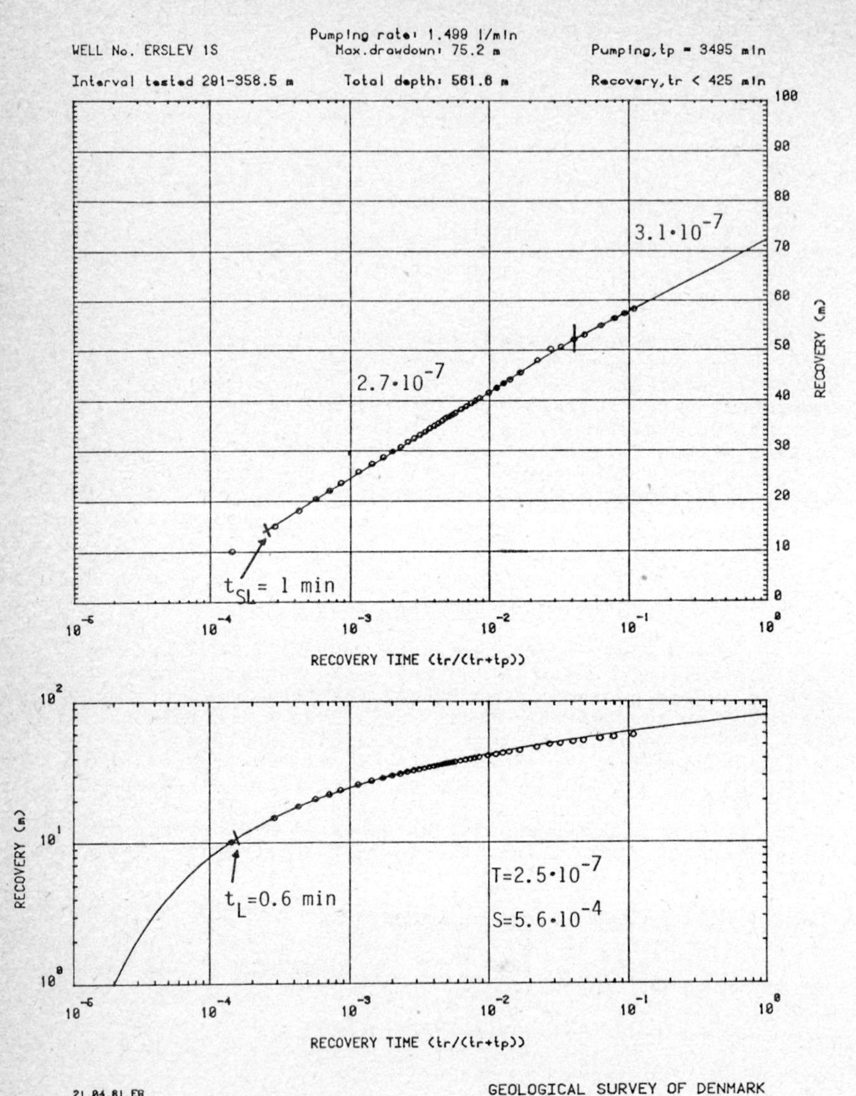

Fig. 2: Recovery, DGU test NO.: E1S-13.

$T=2.5\ 10^{-7}$, $S=5.6\ 10^{-4}$, $k=0.37\ 10^{-8}$, $S_s=8.3\ 10^{-6}$.

Equation 1 may be used to calculate the pressure increase during the recovery if the recovery time is sufficiently small, compared with the pumping time, but more than 1-10 minutes.

When recovery time becomes comparable with pumping time, equation 2 should be used:

$$H = 2.3Q/(4\pi T)\log(t'/(t'+t)) \quad (2)$$

where t = pumping time (sec.)
t' = recovery time (sec.)
H,Q,T = as in eq. 1

One of the assumptions made in eq. 1 and 2 is that Q = constant. This condition was never fulfilled because the Q value depends very strongly on the total lift of the pump and thereby on the drawdown. At the beginning af the pumping, the water had to be lifted from the static water level, i.e. 10-20 m, and the relatively high Q value corresponds to this lift. Then pumping continued, for a very short time, with flow rates much higher than the formation could supply. The storage of the tested interval was practically nil due to the use of packer(s), and rapid drawdown to the maximum value occurred. For this reason only recovery data has been used for calculations.

The most convenient way of calculating the transmissivity of the reservoir is to plot the recovery against a logarithm of time. If the data points can be approximated by a straight line, the transmissivity can be calculated from the slope of that line, using eq. 3:

$$T = 0.183 \times Q/H_{dec} \quad (3)$$

where H_{dec} is drawdown corresponding to one time decade.

The procedure described above applies to both types of plots: $H = H(\log(t'))$ and $H = H(\log(t'/(t+t')))$.

When the data are plotted against the dimensionless time $(t'/(t'+t))$, the static, "undisturbed" pressure in the tested inverval may be estimated by extrapolating the straight line to $t'/(t'+t) = 1$, which corresponds to the infinite recovery time.

Another method of evaluation of the transmissivity (T) and the storage coefficient (S) consists of matching the field data (plotted as logarithm of drawdown or recovery against logarithm of time) with a type curve representing the theoretical solution. The detailed description of the method may be found in /5/, /6/ or /7/. Both methods were used for the evaluation of the pumping tests (Fig. 2-4).

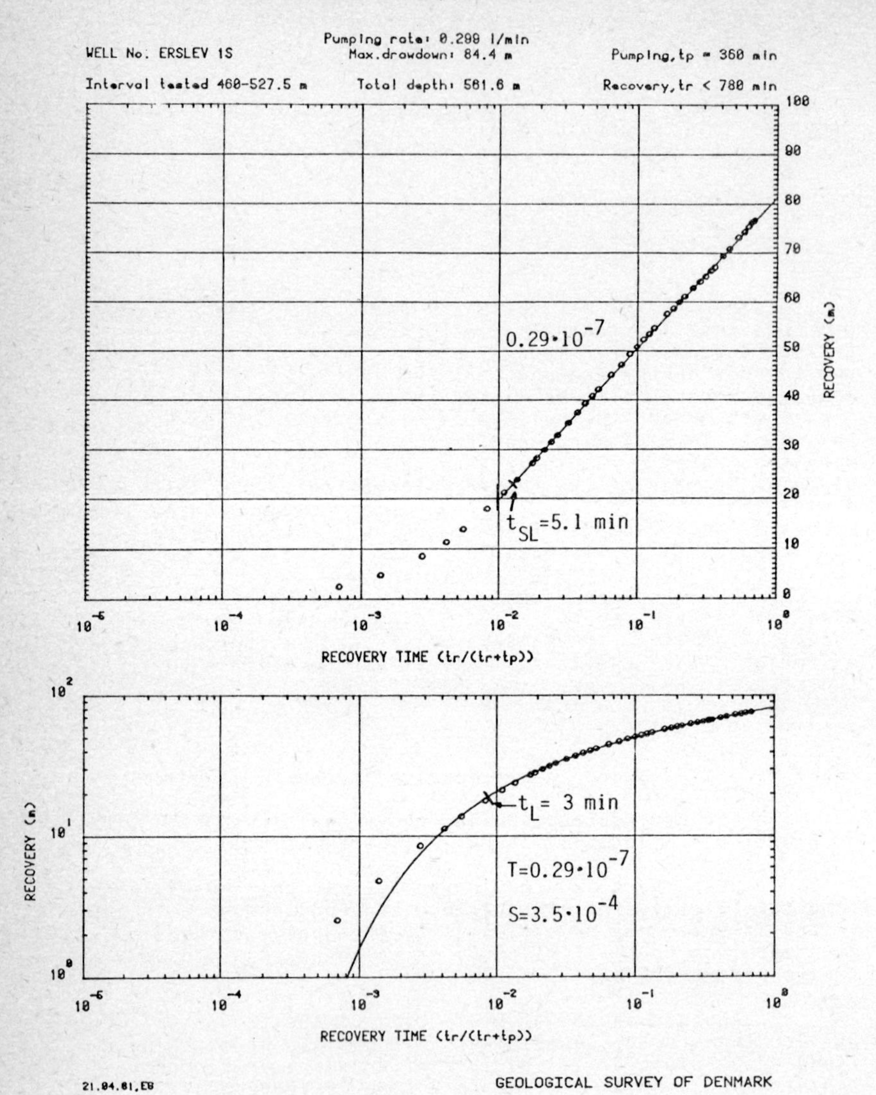

Fig. 3: Recovery, DGU test NO.: E1S-18.

$T=0.29.10^{-7}$, $S=3.5\ 10^{-4}$, $k=0.043.10^{-8}$, $S_S=5.2.10^{-6}$.

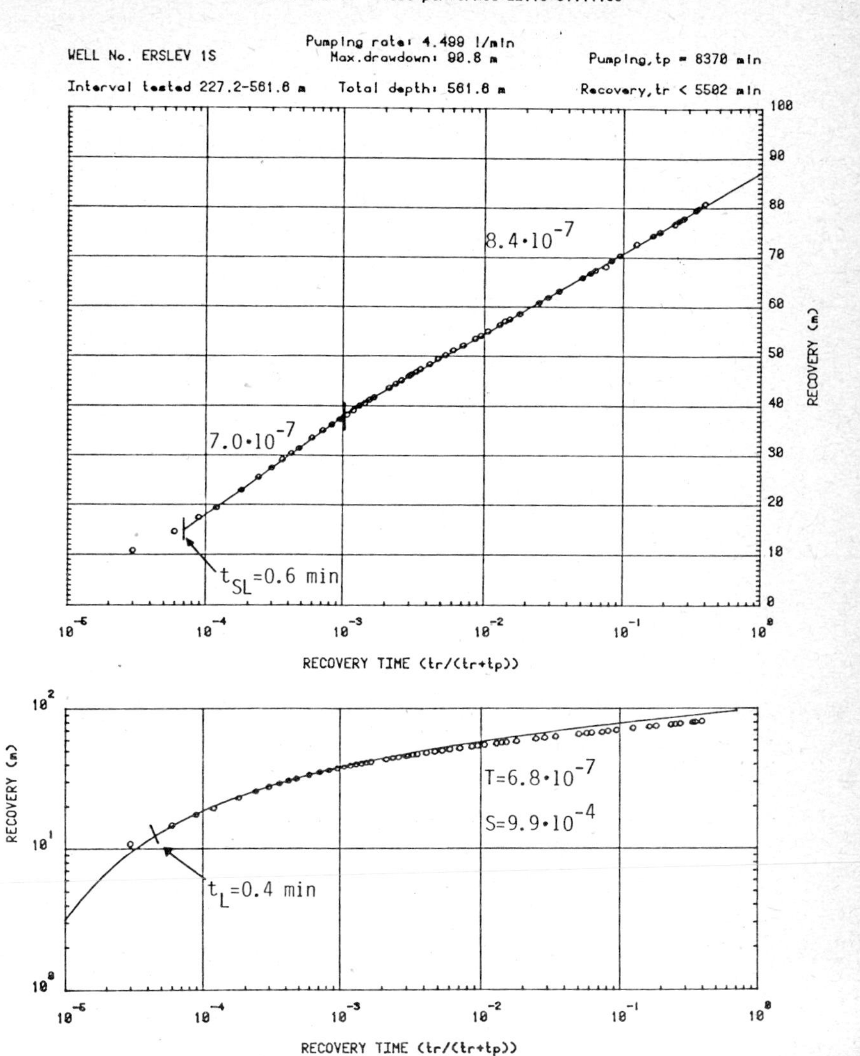

Fig. 4: Recovery, DGU test NO.: E1S-10.

$T=6.8\ 10^{-7}$, $S=9.9\ 10^{-4}$, $k=0.20\ 10^{-8}$, $S_s=3.0\ 10^{-6}$.

This test may be used to control the results of a number of tests performed on the shorter intervals. The transmissivity obtained from this test is about 20% lower than the sum of the transmissivities obtained from the other tests. A presence of highly producing zone in the interval 291-293.5 m bs. may reduce the difference.

2.1.2 Examples

Two examples of recovery, typical for the upper (291-358.5 m b.s.) and the lower (460-527.5 m b.s.) sections of the formation investigated, are shown in Fig. 2 and Fig. 3. Fig. 4 shows recovery data for the interval 227-561.6 m b.s. (TD). The calculations of transmissivity (T) and storage coefficient (S) were performed, using theoretical solutions applicable to infinite, homogeneous and isotropic reservoirs /6/.

The straight-line approximation of the theoretical solution is valid from a certain recovery time, t_{SL}, calculated for every test, using transmissivity and storage coefficient values (T and S) determined for the reservoir.

Another characteristic time value, t_L, has been used for log-log plots. The need for t_L arises from the fact that the radius of the pumping well is not infinitesimal as the theoretical solution requires. Within the bore hole the transmissivity is practically infinite, and as long as $T = \infty$ dominates the rate of the recovery, no theoretical solutions exist. It is (arbitrarily) assumed that the theoretical solution is valid when the radius of influence, R_i, becomes equal or greater than twice the radius of the well. The radius of influence may be defined /7/:

$$R_i = 1.5 \sqrt{T/S \cdot t} \qquad (4)$$

where t is recovery time.

This formula is valid for pumping, but when the superposition principle is applied, R_i may be calculated for early recovery times.

Requiring $R_i = 2r_w$, and taking T and S corresponding to the later part of the data curve, makes it possible to evaluate t_L from equation 4.

2.2 Step-drawdown and step-injection tests

The principles of step-drawdown (SD) and step-injection (SI) tests are similar. A number of successive, constant-rate pumping/constant-rate injections are performed. By manipulating the pumping arrangement, a number of "steps" - quasi steady-state situations with stable drawdown/overpressure and pumping/injection rates - are achieved. The analysis of the corresponding values of drawdowns/overpressures and pumping/injection rates may provide valuable information about fissure permeability.

For fractured reservoirs, the hydraulic conductivity is a function of the pressure changes applied during testing, and k values calculated from injection tests should be higher than k values calculated from drawdown tests.

SI and SD tests performed at Mors gave hydraulic conductivity values within the same order of magnitude as the constant-rate pumping tests.

2.3 Conventional and pressurized slug tests

The conventional slug test (CST) is performed by rapidly adding water to the bore hole. The rate of decrease of the water level in the well depends on the diameter of the well (in the open-hole section), on the diameter of the casing or the stand pipe (from which water is removed), and on the transmissivity of the tested interval.

A transmissivity determination is made by matching field data (the rate of decrease of the head) with type-curves, representing theoretical solutions.

The theoretical solution consists of a family of curves, depending on two dimensionless parameters:

$$\alpha = r_s^2 x S / r_p^2 \quad (4)$$

and

$$\beta = T \times t / r_p^2 \quad (5)$$

where

r_s = radius of the bore hole (m)
r_p = radius of the casing or the stand pipe (m)
S = storage coefficient (dimensionless)
T = transmissivity (m^2/sec.)
t = time (sec.)

T and S may be evaluated from eq. 4 and 5 when α and β are estimated from the type-curve that gives the best fit to the field data.

The method of evaluation of pressurized slug tests (PST) is, in principle, the same as for CST. Two dimensionless parameters, α_1 and β_1, are obtained by matching the field data with type-curves, and the unknowns T and S may be determined.

PST is more suitable for formations of low permeability than CST because the time involved is much shorter. Instead of pouring water into a casing or stand pipe (which was the case during CST), a pressure pulse is applied in the tested interval. The water within the pipe system, from the surface down to the tested interval, is suddenly pressurized by the injection of a small amount of water. If the permeability of the formation is low, this results in a rapid increase of pressure (in this investigation up to 10 bar). The pressure will decline as the extra amount of water added during the pressurization is squeezed into the formation.

The parameters α_1 and β_1 are analogous to α and β (CST), but they are dependent on several additional (known) variables - compressibility, density and the volume of the pressurized water.

Existing tables, giving the theoretical drawdown of pressure as a function of α_1 and β_1, were used to draw type-curves.

The PST gives a rather shallow penetration of the investigated formation and, in the presence of skin, the transmissivity determined will correspond to the invaded zone rather than to the true formation transmissivity.

The theory for evaluation of the slug test was described by J.D. Bredehoeft and S.S. Papadopulos /8/, /9/.

Examples of CST and PST are shown in Fig. 5 and Fig. 6, respectively.

The hydraulic conductivity values obtained from these tests are in reasonable agreement with values obtained from pumping tests, but the values of the specific storage, S_s, are both too great for the conventional and too small for the presurized slug test.

2.4 Labelled slug test

A specfic paper is devoted to the description of the labelled slug test (LST) performed at Mors.

The distribution of hydraulic conductivity obtained from this test is in good agreement with the results of the pumping tests.

3. CONTROL OF THE RESULTS

The investigation of the properties of the low permeable formations gives rise to several problems regarding the conducting of the tests and the evaluation of the results. For this reason it is essential that different testing and calculation methods are used concurrently to determine the investigated parameter - the hydraulic conductivity.

As mentioned before, the results of the constant-rate pumping tests have been controlled by other types of tests: step-drawdown and injection tests, pressurized and conventional slug tests and by the labelled slug test. Furthermore, calculations of the theoretical drawdown were performed, using eq. 1, and a comparison between the calculated and the observed drawdowns were undertaken. For more than 50% of the tests, the deviation from the theoretical solution was less than ±10%, and a deviation greater than ±30% was found for less than 10% of the tests. The difference may indicate either a measuring error or a variation in properties with the distance from the bore hole.

The permeability measurements on the cores follow the same pattern as the hydraulic conductivities determined from the hydraulic tests.

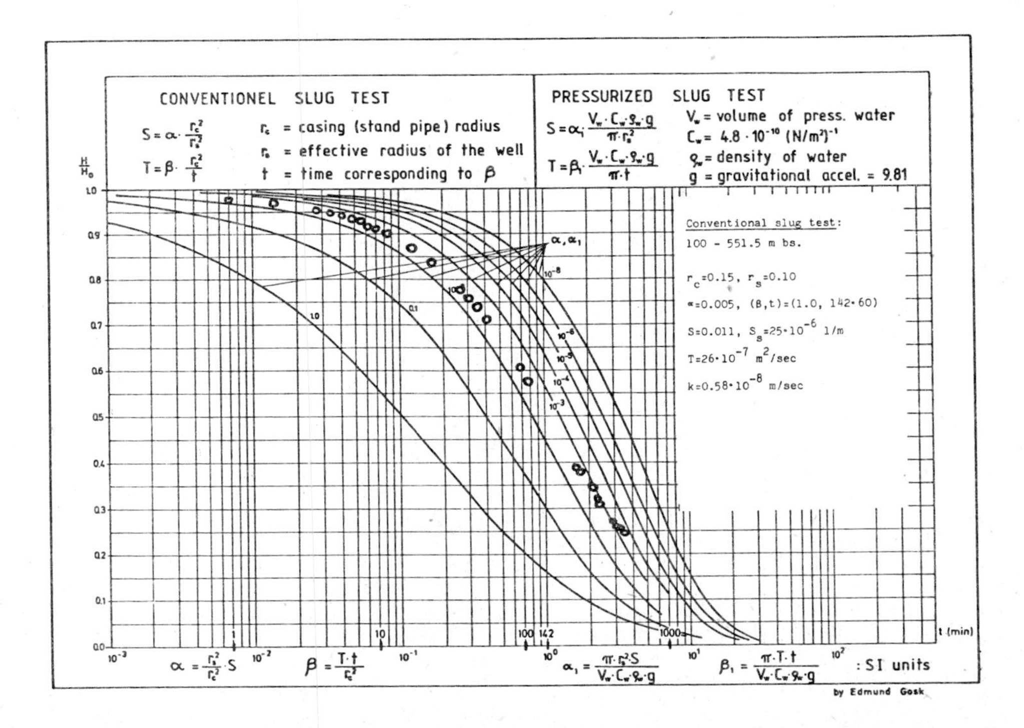

Fig. 5: Example of conventional slug test.

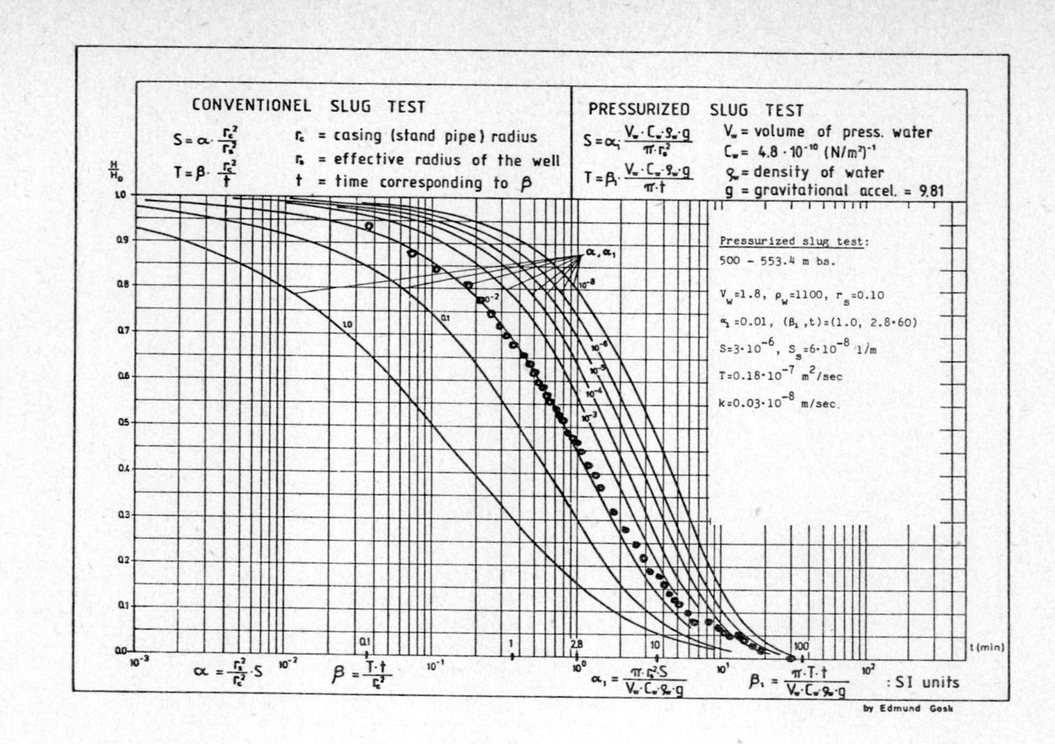

Fig. 6: Example of pressurized slug test.

4. CONCLUSIONS AND RECOMMENDATIONS

1. The design of the test arrangement and the choice of the testing equipment should be considered carefully when testing low permeable formations:

- the use of packers is necessary for pumping tests.

- the use of a wire-line-packer arrangement is recommended as it is quicker, more flexible and simpler than a drill-pipe arrangement.

- the use of a flow or head-controlling system is recommended in order to allow for performance of the tests under well-defined conditions.

2. The testing methods should be chosen in such a way that possible errors, due to equipment failure, can be discovered:

- methods based on different principles and using different technical arrangements are preferable.

- control calculations, which may provide valuable information about the influence of the drilling process on the bore-hole conditions, are advisable.

3. The labelled pumping test may serve two purposes:

- localizing highly-producing and non-producing zones, which then may be investigated by a straddle-packer test.

- determination of the nearly continuous distribution of the hydraulic conductivity at the face of the bore hole.

4. If fissures exist or are expected, pumping tests should be supplied with injection tests.

5. REFERENCES

/1/ Gosk, E., N. Bull and L.J. Andersen: "Hydrogeological Programme, Mors Salt Dome. Hydrogeological Main Report. Geological Survey of Denmark, Copenhagen, April 1981. 192 pages and 98 figures (Internal report).

/2/ Joshi, A.V.: "Results of Geological Investigations of the Mors Salt Dome for The Disposal of High-Level Radioactive Waste in Denmark", Workshop on SITTING OF RADIOACTIVE WASTE REPOSITORIES IN GEOLOGICAL FORMATIONS, Paris 19-22 May, 1981.

/3/ Andersen, L.J., L. Byrne and P. Frykman: "Results from Impression Packer Technique for Location of Fissures in Boreholes in White Chalk, Mors, Denmark", Workshop on SITTING OF RADIOACTIVE WASTE REPOSITORIES IN GEOLOGICAL FORMATIONS, Paris 19-22 May, 1981.

/4/ Andersen, L.J., N. Bull and E. Gosk: "Results of a Labelled Slug Test in a Low Permeability Formation of White Chalk, Mors, Denmark",
Workshop on SITTING OF RADIOACTIVE WASTE REPOSITORIES IN GEOLOGICAL FORMATIONS, Paris 19-22 May, 1981.

/5/ Kruseman, G.P. and N.A. de Ridder: Analysis and Evaluation of Pumping Test Data", Bulletin 11, International Institute for Land Reclamation and Improvement, Wageningen, The Netherlands 1976.

/6/ Theis, C.V.: "The relation between lowering of the piezometric surface and the rate and duration of discharge of a well using ground water storage", Trans. Am. Geophys. Un., 16th annual meeting, 1935.

/7/ Bear, J.: "Hydraulic of Groundwater", McGraw-Hill Series in Water Resources and Environmental Engineering, 1979.

/8/ Bredehoeft, J.D. & Papadopulos, S.S.: "A Method for determining the hydraulic properties of tight formations". Water Resources Research, Vol. 16, No. 1, 1980.

/9/ Papadopulos, S.S. et.al., 1972: "On the analysis of "Slug Test" data. Water Resources Research, Vol. 9, No. 4.

DISCUSSION

A.V. JOSHI, Denmark

From the analysis of pumping results (incl. injection tests) can you make any observations regarding fissure permeability ?

E. GOSK, Denmark

We cannot conclude about the presence of the fissures from the results of the tests performed. It is mainly due to the rather poor quality of the injection tests (no special equipment was available). However, the pumping tests do not indicate any fissure permeability and the K-evaluation from the tests correspond to the K-values obtained from laboratory tests (matrix permeability).

A.V. JOSHI, Denmark

What is the reason for the difference in permeability between Maastrichtian and Campanian ? Is it because of the difference in the physical properties of the chalk ? How does this agree with your theory of a compression zone ?

E. GOSK, Denmark

The decreasing permeability correlates to the decreasing porosity. Lothology and hardness of the rock are different for the Maastrichtian and Campanian which may as well influence the permeability.

The discussion about the compression theory is not included in the paper and I feel that introducing this subject now will require more time than it is available.

J.D. MATHER, United Kingdom

In your paper you emphasised the importance of using a number of different techniques in hydraulic testing. Were you able to use groundwater chemistry and age-dating techniques as additional checks in the conclusions you came to using hydraulic testing ?

E. GOSK, Denmark

Unfortunately we did not. The main reason for that was the combination of : the low flow rate, the relatively large diameter of the borehole and the limited time available for the tests, which made it difficult to obtain the representative water samples. The not very well defined quality of the liquid within the borehole at the beginning of the tests made it difficult to determine the formation water composition from the mixtures obtained during tests. We still have some water samples which may be analyzed later if we found it feasible.

N. VANDENBERGHE, Belgium

For what reason did you prefer an ejector pump over submersible pumps ?

E. GOSK, Denmark

The ejector pump is very flexible with regard to the combinations of drawdown and flow rates. An alternative to ejector pump would be a number of submersible pumps which had to be tried on the tested intervals. Going in -and out of the hole (if the wrong pump is chosen) is time consuming and costly.

Furthermore, the ejector pump could be easily incorporated in the testing arrangement.

RESULTS FROM IMPRESSION PACKER TECHNIQUE FOR LOCATION OF FISSURES IN BOREHOLES IN WHITE CHALK, MORS, DENMARK

Authors: L.J. Andersen, L. Byrne and P. Frykman,
The Geological Survey of Denmark
Copenhagen, Denmark

ABSTRACT

In connection with site investigations to evaluate the possibility of geological disposal of radioactive waste in a salt dome a Hydrogeological Programme was carried out. A number of bore holes were drilled to depths of 550 m. Geological and hydrogeological investigations were performed.

The hydrogeological tests were carried out as packer tests, bottomhole tests using one packer, or straddle-packer test using two packers.

In connection with these tests a rubber sleeve was installed over the packer and sidewall impressions of the hole were taken. Irregularities such as faults and fissures and possibly solution features were localized.

The paper describes the method, procedure, equipment and the results achieved.

INTRODUCTION

Impression packer technique has been applied in Upper Cretaceous chalk and limestone formations at the Mors salt dome, Denmark (Figure 1). The technique was used in connection with an investigation programme performed to evaluate a future possibility for disposal of high-level radioactive waste in rock-salt formations in Denmark.

The investigation programme includes two 3500 m deep test wells into the rock-salt formations and four hydrogeological test wells in the overlying chalk and limestone formations to a depth of 560 m b.s. The technique was introduced to obtain supplementary information about the fissure distribution and dimensions. The general description of the investigation programme is given by A.V. Joshi [1].

IMPRESSION-PACKER TECHNIQUE

In principle the impression-packer technique consists in applying an impression rubber sleeve to a packer, lower the packer to the depth of interest and inflate the packer. With a sufficient pressure, the rubber will flow into irregularities of the side wall of the bore hole.

By deflation of the packer, the impression sleeve is detached from the side wall with the impressions obtained. After retrieval, the packer should be reinflated to the hole diameter, and the impression sleeve photographed and possibly oriented.

The impression-packer technique and the procedure are described in detail by Lynes [2]. Commercial impression rubber sleeves are available from Lynes Petro Tech Limited.

The impression rubber is a rubber with little or no elasticity and will record all irregularities of the borehole side wall, i.e. rock textures and fractures, faults and solution cavities as well as tool marks and impressions of cuttings. A mud cake on the side wall may prevent impressions of the features below it. However, the presence of a mud cake in itself indicates some matrix permeability.

THE MORS SITE INVESTIGATION

Geology

The Mors salt dome consists of Zechstein salt deposits with the cap rock at a depth of approximately 700 m b.s. In the investigated area the Upper Cretaceous sequence from 560 to 60 m b.s. consists of Maastrichtian white chalk, soft, friable with some beds and nodules of flint overlying the Campanian and Santonian limestone, a white to light grey, moderately hard to hard limestone with marly beds. The sequence has been fractured and faulted due to halokinetic movements of the rock-salt formation below.

The Cretaceous sequence is superposed by Quaternary deposits of meltwater sediments and clayey till.

FIGURE 1.
Location of the investigated area on Mors, Denmark.

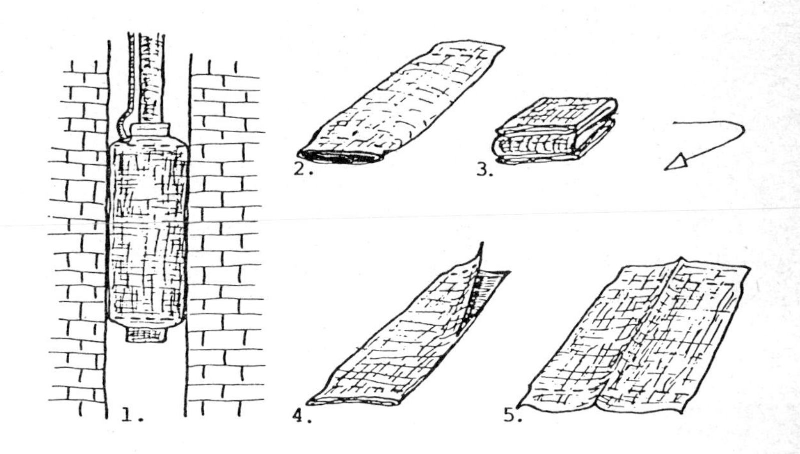

FIGURE 2.
Route diagram for impression sleeve.
1. packer with rubber sleeve positioned in the bore hole.
2. sleeve removed from packer, 3. rolled for transport.
4. cut open and 5. laid flat for inspection.

Packer arrangement

In the actual investigation a Lynes SCI PIP packer (Surface Controled Inflatable Production Injection Packer) 5 5/8 inch. OD. was used. The packer was mounted on a wireline arrangement and inflated with water pumped from a small high-pressure hand pump.

The rubber used for the impression sleeves is a non-vulcanized rubber tube normally used for fire-hose manufacture. The rubber sleeve is pulled over the deflated packer, and fixed to the packer by banding straps at top and bottom.

The applied inflation pressures have varied from 40 to 80 bars, and maintained from one to two days during the pumping-test period. However, pressures of 5-10 bars maintained for a few hours will presumably be sufficient to record even fine details of irregularities of the side wall. The borehole diameter varies from 9 to 12 inches, and the packer diameter (deflated) is 5 5/8 inches.

Data description and presentation

After being photographed the sleeve is removed from the packer and brought to the laboratory for detailed study and description.

In the laboratory the sleeve is cut along one side, opened out and laid flat to facilitate inspection (Figure 2).

The impression observed on the sleeve is drawn on a diagram as shown in Figure 3 and 4. The different symbols used to describe the impressions are explained in the figures.

INTERPRETATION OF IMPRESSION DATA

In total 10 impression packer sleeves have been investigated and described in the laboratory. Two examples are shown in Figure 3 and 4, [3] and [4].

The impressions recorded are dominated by different types of lines and ridges, curved as well as straight, and impressions of areas distinguished by a different surface roughness or delineated by lines around the areas.

On the impression packer sleeve shown in Figure 3 areas of two types are recorded. One type is slightly rounded in outline and smooth on the surface, the other forms a nearly horizontal layer and has a very irregular relief with sharp edges.

Both types of areas are interpreted as impressions of flint. The areas with smooth surface are interpreted as flint nodules apparently broken in one blow. The irregular horizontal layer has apparently been drilled more slowly, resulting in a smashing of the flint layer.

From the core material it is evident that flint nodules occur in the investigated interval.

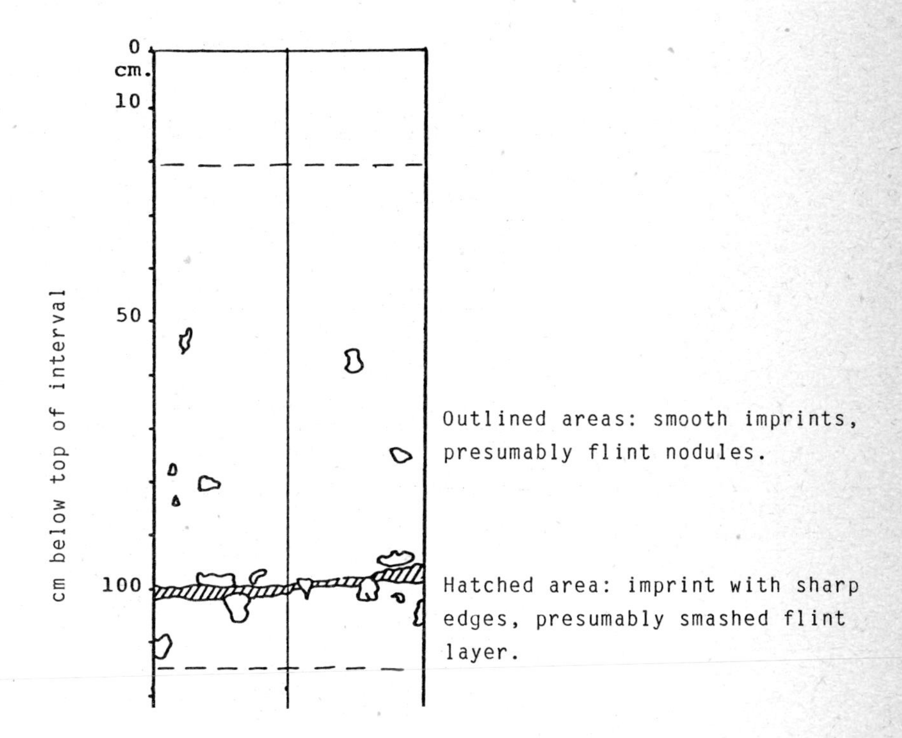

FIGURE 3.
Impression diagram with flint layer and flint nodules from lo7.65-lo9.oo m b.s. in well E3S, Mors, Denmark.
The dashed lines indicate limits for the packed interval.

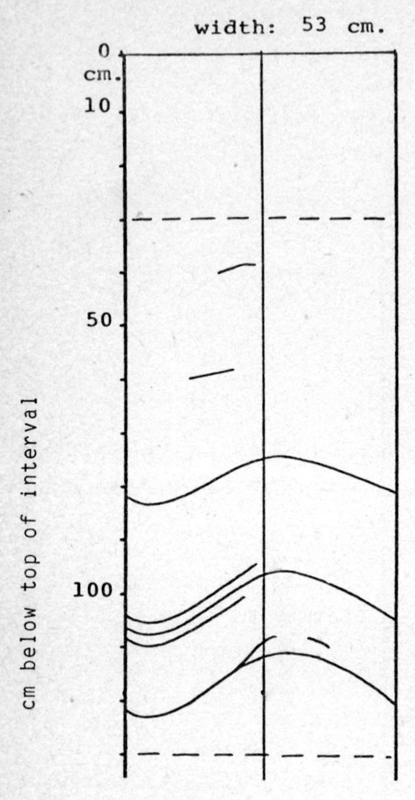

Faint line, curved, positive relief. A=11 cm; T=32 cm; B=7 cm.

Clear lines, curved, positive relief; lower one apparently branched, A=13 cm; T=3o cm; B=6,5 cm.

FIGURE 4.
Impression diagram with inclined fracture planes at 273.3o-275.1o in well E2S, Mors, Denmark.
The dashed lines indicate limits for the packed interval.

Legend:

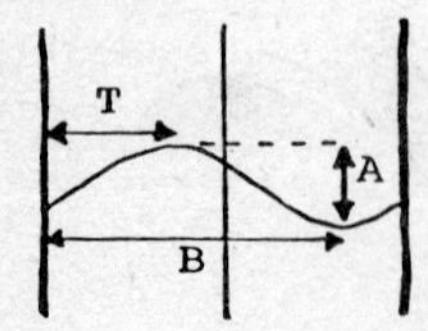

A: amplitude of curved line.
B: distance from crest to left side of packer sleeve.
T: distance from trough to left side of packer sleeve.

Thin planar structures as fractures, faults and clay seams intersecting the bore hole will show up on the impression-packer sleeve as thin ridges with a positive relief. The ridges observed on the sleeves range in size from hairline to a width of a couple of millimetres and a height of over one millimetre.

When these planar features intersect the bore hole at an angle, the lines on the impression sleeve will have a sinusoidal form. When the features are normal to the bore hole, the lines are horizontal.

The impression diagram shown in Fig. 4 shows several curved lines with approximately the same orientation, and a calculated dip of approx. 20°. The lines are faint to clearly distinguishable and with positive relief. One of the lines seems to have a short branch above the main line.

It is possible that these features represent solution seams where the thin residual layer of weaker clay has been removed by the drilling process, leaving an invagination on the side wall. However, in this interval it is known from core material that some solution seams have a stylolitic configuration, and this sutured outline should also be observed on the impression if it had been caused by casting of a solution seam. Therefore, until further evidence is available, it is considered most likely that the lines on the impression sleeve indicate fracture planes or faults and not solution seams.

Vertical lines with positive relief have been interpreted as tool marks, made when tools have been run in and out of the hole. They have not been interpreted as vertical fractures as it was thought that they would have a higher relief and better definition if this had been the case. On some sleeves concentric, nearly horizontal lines can be seen. These are thought to have been caused by the rotary drilling process. Some sleeves, notably those from shallower depths, were seen to be covered with a chalky mud cake. This may cause features not to be recorded on the rubber. However, this in itself would indicate a degree of permeability where the packer had been set. From pumping tests it appears that the upper part of the penetrated sequence has a higher permeability.

CONCLUSIONS

Experiences with the impression-packer technique from the Mors site investigations show that impressions of fractures and other features can be obtained from even relatively soft formations such as the chalk. In more hardened lithologies it is obvious that fissures will give well-defined impressions and important information about fissure distributions and dimensions.

The impression-packer technique implies a less complicated procedure and equipment for obtaining information from the bore hole than conventional coring. It can be carried out on a wire-line arrangement without a drilling rig.

By introduction of an orientation device the true spatial distribution of the features might be obtained, e.g. dip and strike of fault planes.

In low permeable formations it might be possible by a modified technique in connection with a straddle-packer arrangement to obtain impressions from large intervals of the bore hole.

REFERENCES

[1] Joshi, A.V.: "Result of geological investigations of the Mors salt dome for the disposal of high-level radioactive waste in Denmark", Proc. OECD Workshop on siting of Radioactive Waste Repositories in Geological Formation. Paris 19-22 May, 1981.

[2] Lynes: "Technical Manual", Lynes Impression Packer Product No. 300-07, Febr. 4, 1977, pp. 1-7.

[3] Knudsen, J.: "Geological Well Completion Report Hydrogeological Well Erslev 3 S, The Mors Salt Dome", Geological Survey of Denmark, Copenhagen, March, 1981, Internal Report.

[4] Byrne, L.: "Geological Well Completion Report. Hydrogeological Well Erslev 2 S, the Mors Salt Dome", the Geological Survey of Denmark, Copenhagen, March, 1981, Internal Report.

DISCUSSION

F. GERA, Italy

Could you compare the impression packer technique, from the points of view of performance and cost, with other methods for investigating the hole walls, such as the downhole television or the acoustic televiewer ?

L.J. ANDERSEN, Denmark

In the way we used this technique, as an additional information in connection with the pumping test, the cost is very low, just the price of the rubber sleeve and the mounting on the packer. In case of making an impression packer test alone, the cost is equal to the cost of setting and releasing the packer. I would think that it should be the cheapest of all methods.

RESULTS OF A LABELLED SLUG TEST IN A LOW-PERMEABILITY FORMATION OF WHITE CHALK, MORS , DENMARK

L.J. Andersen, N. Bull and E. Gosk,
The Geological Survey of Denmark,
Copenhagen, Denmark

ABSTRACT

In connection with the Hydrogeological Programme carried out on the Mors salt dome, Denmark, a Labelled-Slug-Testing method was used to determine the vertical distribution of the hydraulic conductivity of the White Chalk formation in an open bore hole, 550 m deep.

An attempt was made to determine the hydraulic conductivity of - and its distribution over - the tested interval; a measurement of the radioactive pulse velocity has been taken; and the results have been compared with the results from packer-pumping tests, in order to control the reliability of the method.

The equipment used, the results of the test and method is described and discussed.

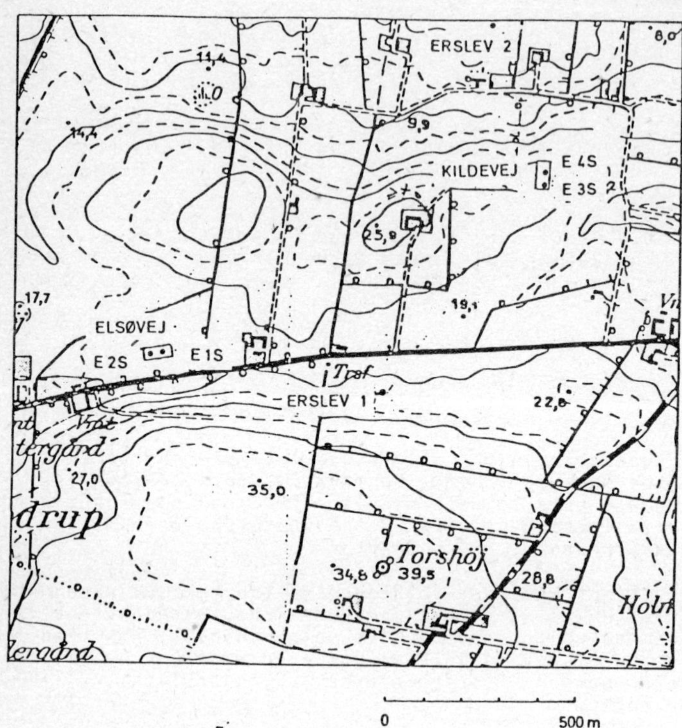

Fig. 1 .: Location of the two well sites Elsøvej and Kildevej, with the four hydrogeological wells and the deep test holes Erslev 1 and Erslev 2, Mors salt dome. Denmark.

1. INTRODUCTION

The Ground-water movement through chalk/limestone rocks is believed to take place partly through the rock matrix and partly through fractures and fissures, which have been created after deposition of the rocks. The knowledge of matrix and fracture permeability has great importance in connection with the investigation of ground-water use, waste disposal (chemical, radioactive), oil and gas production etc. In rocks with a high hydraulic conductivity due to either matrix or fractures it is possible to make the hydraulic conductivity calculation on the basis of conventional pumping tests, impeller flowmeter logs and a series of bore-hole logs (resistivity, nuclear, acoustic and caliper). In rocks with a very low hydraulic conductivity the conventional pumping test will be time-consuming, and the impeller-flowmeter log cannot be run due to a mechanical minimum speed for the impeller. The bore-hole logs are able to produce information about hydraulic properties, but it is an indirect method, which has to be proved by other kinds of tests.

A quick and accurate method to evaluate the hydraulic conductivity in low conductive rocks is the labelled slug test, which consists of four steps:

1) Placing of a radioactive tracer at several levels in the bore hole
2) Pumping on the bore hole with constant capacity
3) Periodic logging of the time-depending depth to the single tracer pulses during the pumping
4) Determination of the pulse-velocity distribution within the bore hole.

The velocity of the pulses as a function of depth and time defines the hydraulic conductivity and its vertical distribution, and at the same time gives information about the presence of fissures. This procedure has earlier been used by Marine /1/ for location of fissures in a low permeable formation. However, the authors have developed the method further for determination of the hydraulic conductivity of the penetrated formations.

2. GEOLOGY AND HYDROGEOLOGY OF THE INVESTIGATION SITE

The area studied (Fig. 1) is situated at Mors, Northern Jutland, on the top of a Permian salt diapir, which has uplifted a younger formation of the Cretaceous and the Tertiary age. A number of wells have been drilled, two down into the salt diapir itself, and four to a depth (550 m-560 m b.s.) just above the cap rock. Information from the bore holes and some seismic and geoelectrical investigations in the area show that there is about 60 m of Quaternary clay, sand and gravel, and below this there is Upper Cretaceous formation consisting of chalk and limestone. The upper part of the chalk formation is white, soft and friable. Below 350-400 m b.s. the hardness increases, and the porosity decreases from about 40% to about 15% at the bottom, *[2]*.

LABELLED SLUG TEST. WELL ERSLEV NO. 1S.

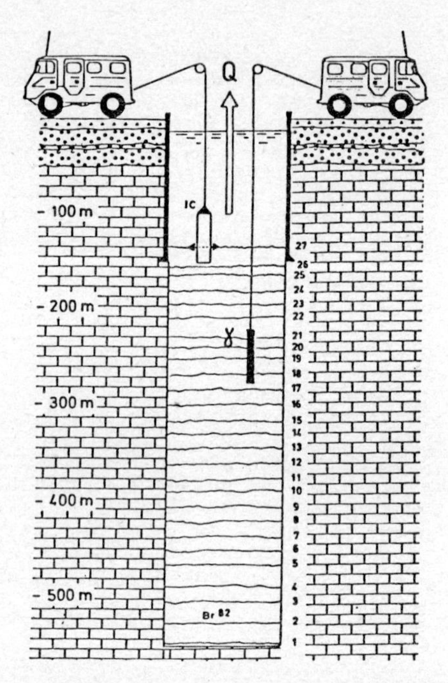

Fig.: 2. Labelled slug test. Technical arrangement

1 to 26: Labelled slug. Isotope tracer Br^{82}.

γ: DGU Gammalogging probe.

IC: IC injection tool.

GEOLOGICAL SURVEY OF DENMARK

APRIL 1981 NILS BULL

The labelled slug test has been made in one of the shallow wells, where the core inspection and a series of bore-hole logging indicate numerous fractures in several intervals. The pumping tests performed gave permeability values corresponding to the permeability of the matrix determined on cores in the laboratory.

3. THE LABELLED SLUG TEST

The test was performed from 20th to 22nd January, 1981, with a total pumping period of about 33 hours, *[2]*. The Danish Isotope Centre, Copenhagen (IC) delivered the tracer, Bromine-82 solution, and designed and performed the piston pump probe used to set the pulses in the well. The technical arrangement is shown in Fig. 2.

The piston pump probe was filled with a Bromine-82 solution, mounted on a logging cable and lowered to the bottom of the bore hole.

27 Bromine-82 pulses were released into the well in the interval 550 m to 130 m below ground level (Fig. 2). A gamma log after the pulse setting showed that 23 of the 27 labelled slugs have been recognizable (Fig. 3). Two pulses seem to have been misplaced in the interval between 520 m and 550 m b.s. The bore hole has not been cleaned for mud after the total drilling depth has been reached, and both the caliper and the mud-resistivity log indicate heavy mud at the interval 520 m to 550 m b.s. Because of this, the pulse setting was incorrect within this interval. Two other pulses in the depth of 355 m and 340 m b.s. could not be recognized on the logs.

To produce a water flow from the formation into the bore hole a submersible pump was mounted 96 m b.s. The pumping was started on 20th January, 1981, at 23 hours, and ended on 22nd January at 8.30 o'clock. The position of the single pulses during the 33 hours of pumping has been measured by 14 gamma logs (Fig. 3 and Table I).

The gamma logs were run without a centralizing device, which means that there could be expected some measuring inaccuracy according to placing of the probe in the labelled slug-flow geometry. The low air temperatures during the testing period gave rise to ice formation on the cable and on the depth-measuring unit, and for that reason some inaccuracy in the exact definition of the depth to the single labelled slugs occurred. Due to this and the very low flow velocity, less than 1 m/hr. in depths from 375 m to 550 m b.s., it has been necessary to determine the single pulse velocities by averaging over a number of gamma logs (Fig. 4). The measuring point of the single labelled slug has been either the maximum of the pulse or an abrupt radiation change below or above the maximum of the pulse.

LABELLED SLUG TEST. ERSLEV NO. 1S.

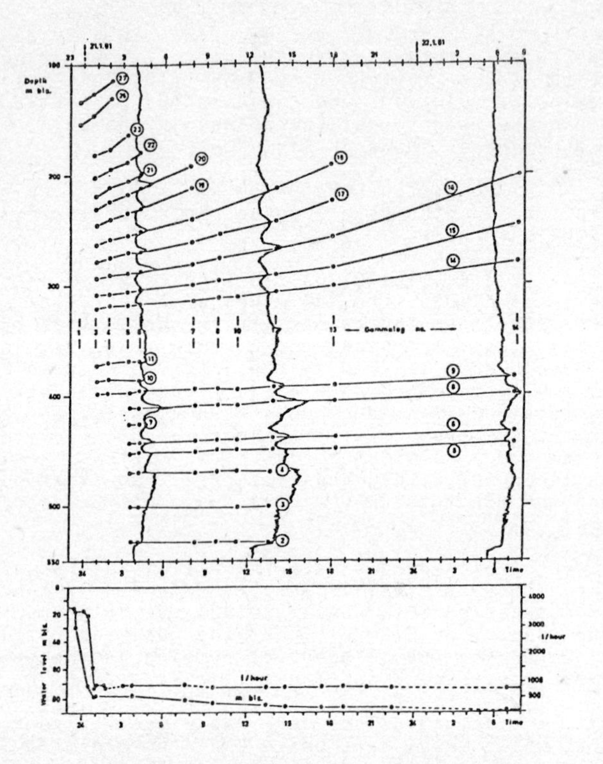

Fig.: 3. Labelled slug ⑤ position by time and depth. Variation in pumping capacity and water level drawdown during pumping on the well from 20.1.81. to 22.1.81.

GEOLOGICAL SURVEY OF DENMARK

APRIL 1981 NILS BULL

Labelled slug test.   Well Erslev no. 1 S

	20/1 1981			21/1 1981											22/1 1981
27	132.8 19^{22}	132.8 23^{44}	132.0 23^{55}	126.2 0^{40}											
26	152.9 19^{24}	152.9 23^{44}	152.0 23^{55}	145.4 0^{41}											
25	165.7 19^{25}														
24	178.7 19^{27}														
23	192.7 19^{30}			185.8 0^{46}	175.7 1^{59}	164.4 3^{10}									
22	208.0 19^{33}			201.5 0^{48}	194.1 1^{58}	187.5 3^{11}	179.0 4^{13}								
21	224.7 19^{35}			217.7 0^{50}	211.6 1^{56}	205.3 3^{12}	199.2 4^{13}								
20	237.1 19^{38}			231.2 0^{51}	224.1 1^{55}	217.5 3^{14}	211.9 4^{13}	192.0 7^{48}							
19	249.7 19^{41}			245.6 0^{52}	240.1 1^{54}	235.0 3^{15}	230.1 4^{12}	212.0 7^{50}							
18	266.1 19^{44}			262.7 0^{54}	259.3 1^{51}	255.0 3^{17}	252.0 4^{11}	240.0 7^{53}					211.6	190.5 17^{53}	
17	280.7 19^{46}			278.1 0^{56}	275.1 1^{50}	271.8 3^{19}	270.0 4^{11}	260.3 7^{57}	254.7				242.5	223.7 17^{57}	
16	295.5 19^{49}			293.4 0^{57}	291.2 1^{49}	289.2 3^{20}	287.1 4^{08}	280.1 7^{58}	275.3 9^{54}				267.6	257.3	201.2 7^{38}
15	312.0 19^{52}			308.8 0^{59}	307.7 1^{47}	306.3 3^{21}	304.3 4^{08}	299.5 8^{01}	295.7 9^{52}				290.5	284.5	245.9 7^{36}
14	325.1 19^{55}			321.9 1^{00}	319.4 1^{46}	318.8 3^{22}	317.4 4^{08}	313.5 8^{02}	312.0 9^{51}		310.3		306.8	300.5	279.9 7^{31}
13															
12															
11	372.7 20^{00}			372.3 1^{06}	370.7 1^{42}	369.8 3^{29}	369.3 4^{08}								
10	388.1 20^{04}			387.2 1^{06}		386.8 3^{29}									
9	399.3 20^{06}			398.8 1^{09}	397.8 1^{40}	397.5 3^{31}	395.2 4^{07}	394.0 8^{07}	393.4 9^{47}		394.1 11^{12}		392.4 13^{42}	391.2 18^{12}	384.9 7^{21}
8	414.3 20^{09}					412.0 3^{32}	411.0 4^{06}	409.8 8^{08}	408.1 9^{46}		408.4 11^{13}		406.4 13^{42}	405.7 18^{12}	399.6 7^{21}
7	429.2 20^{12}														
6	445.2 20^{16}					443.8 3^{36}	442,8 4^{05}	441.6 8^{13}	440.1 9^{43}		440.6 11^{15}		438.6 13^{39}	438.0 18^{16}	433.5 7^{16}
5	459.0 20^{20}					452.9 3^{41}	452.1 4^{03}	451.6 8^{16}	449.0 9^{42}		450.0 11^{16}		449.7 13^{35}	448.0 18^{17}	443.8 7^{14}
4	472.3 20^{30}					470.3 3^{41}	469.9 4^{03}	469.1 8^{18}	468.1 9^{39}		468.7				
3									501.4			501.9 11^{23}			
2	533.0 20^{34}								532.5			533.1 11^{26}			
1	544.8 20^{36}														
	0	1	2	3	4	5	6	7	8	9	10	11	12	13	14

Table I :   Labelled slug position by time and depth.

1 - 27: Labelled slug no. 01 to 27. 0 - 14: Gamma log no. 0 to 14.

544.8 20^{36}: Labelled slug 544.8 m b.s. at 20^{36} hour.

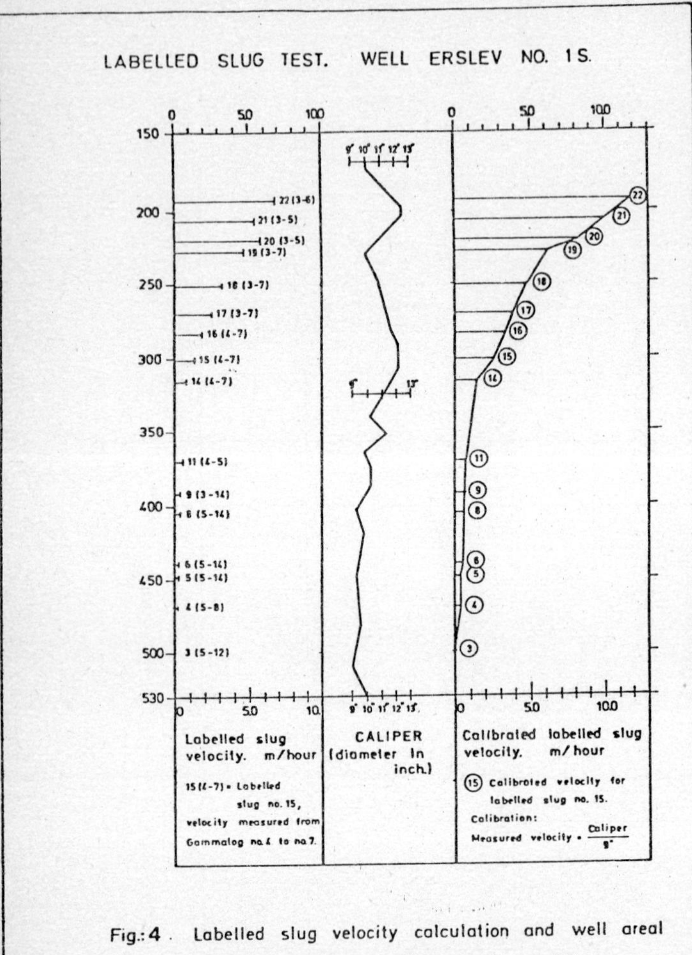

Fig.: 4 . Labelled slug velocity calculation and well areal change correction of the velocity profile.

GEOLOGICAL SURVEY OF DENMARK

APRIL 1981 NILS BULL

The velocities within the single bore-hole intervals (Fig. 4) were corrected for the variation of the cross-sectional area of the bore hole. All the velocities have been calibrated to a theoretical bore-hole diameter of 9 inches. A comparison of the measured and the calibrated pulse-velocity profiles shows that the intervals with an apparent decrease of the measured velocities upwards have been changed to an expected upward increase in calibrated velocities.

4. HYDRAULIC CONDUCTIVITY DETERMINATION FROM THE LABELLED SLUG TEST (LST)

The determination of the average hydraulic conductivity from LST was calculated on intervals defined by the position of the radioactive slug during different runs with the logging tool. Furthermore, averaging over intervals corresponding to the intervals investigated by pumping tests is undertaken for comparison of these two testing methods.

Assuming:

1. that the open section of the bore hole may be divided into a number of intervals for which homogeneous and isotropic conditions exist,

2. that the intervals are defined by the velocity profile obtained from the positions of the radioactive pulses, Fig. 4,

3. that only horizontal flow takes place,

4. that the drawdown is identical for all of the intervals, and

5. that the flow rate may be regarded as constant during the pumping period,

the drawdown (H), the flow rate (Q), the pumping time (t), the hydraulic conductivity (k), and the specific storage (S_s) are related by eq. 1, [3].

$$H = Q/(4\pi k \cdot l) \ln(2.25 \cdot k \cdot t/(r_w^2 \cdot S_s)) \qquad (1)$$

where l is the length of the interval, and r_w is the radius of the well.

This equation is an approximation to the exact solution but for pumping times involved the error is minimal.

When the five above-mentioned assumptions are valid, eq. 1 may be applied for each single interval:

$$H = Q_i/(4\pi k_i \cdot l_i) \ln(2.25 \cdot k_i \cdot t/(r_w^2 \cdot S_s)) \qquad (2)$$

where Q_i is the inflow from the interval l_i.

Q_i was determined from the values of the calibrated velocity measured as velocities of the radioactive pulses. The flow in the bore hole is laminar, but because of the lack of the centralizing device for the pulse probe no correction for the shape of the velocity profile was made, and it is assumed that the pulse velocity represents the average velocity in the bore hole.

The principle of the calculation of the Q_i values is visualized in Fig. 5.

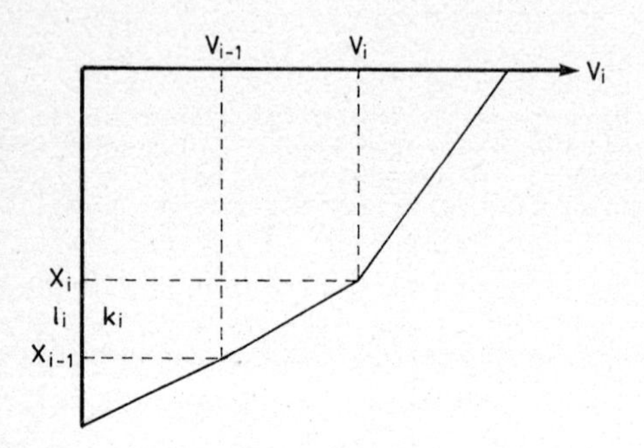

Fig. 5. Principle of calculation
V_i indicates the measured velocity at the end of the interval i;
k_i is the constant hydraulic conductivity interval l_i

Not all of the 27 pulses could be used for calculations. In Table II the calibrated velocities for 22 pulses are calculated. The inflow from the "i"-th interval, Q_i, is calculated, using eq. 3:

$$Q_i = r_w^2 (V_i - V_{i-1}) \quad (3)$$

combing eqs. 2 and 3 gives

$$H = r_w^2 (V_i - V_{i-1})/(4\pi k_i \cdot l_i) \ln(2.25 \cdot k_i \cdot t/(r_w^2 \cdot S_s)) \quad (4)$$

Equation 4 is used for the calculation of the average hydraulic conductivities corresponding to the intervals defined by the velocity distribution calculated from the positions of the single pulses. The only unknown in eq. 4 is the value of the hydraulic conductivity if an estimation of the specific storage (S_s) is made. All the other variables are determined from the labelled slug test, Table II.

The average hydraulic conductivities corresponding to the single intervals may be calculated, using values from Table II and H = 70 m, $S_s = 5\ 10^{-6}\ m^{-1}$ and $r_w = 0.114$ m.

Using the above-mentioned values and rearranging eq. 4 yields

$$k_i = 4.64 \cdot 10^{-5} (V_i - V_{i-1}) / l_i \cdot \ln(3.46 \cdot 10^7 \cdot k_i \cdot t_i) \qquad (5)$$

The velocity and time values from Table II were converted from practical units to SI-units before use in eq. 5.

Interv. No.	Interv. m b.s.	Interval length (l_i) m	Time, t_i hrs.	Velocity, V_i m/hr.	Hydraulic conductivity k_i 10^8, m/sec.
1	501.0-469.2	31.8	7 1/2	0.4	0.048
2	469.2-448.4	20.8	7 1/2	0.4	0.0
3	448.4-438.7	9.7	7 1/2	0.45	0.039
4	438.7-405.8	32.9	15	0.55	0.024
5	405.8-391.9	13.9	15	0.65	0.066
6	391.9-370.0	21.9	6	0.77	0.040
7	370.0-315.7	54.3	4 1/2	1.54	0.12
8	315.7-301.7	14.0	5	2.73	0.95
9	301.7-283.3	18.4	5	3.50	0.43
10	283.3-269.2	14.1	5	4.08	0.42
11	269.2-251.4	17.8	5 1/2	4.85	0.45
12	251.4-228.8	22.6	5 1/2	6.26	0.68
13	228.8-221.6	7.2	5 1/2	8.44	4.0
14	221.6-208.5	13.1	3	10.12	1.4
15	208.5-190.3	18.2	3	11.86	1.0

Table II: Results of the labelled slug test
V_i = velocity calibrated to r_w = 0.114 m (Ø = 9")
t = time corresponding to V_i

In order to make a comparison of the results of the LST and the pumping tests the k values have been averaged over the corresponding intervals in Table III, and the results are visualized in Fig. 6.

From comparison of the results obtained from LST and from the pumping test, it may be concluded that the agreement is satisfactory, particularily if the accuracy, which may have been achieved for both methods is taken into account.

	LABELLED SLUG TEST		PUMPING TEST
Interval m b.s.	k_i 10^8 m/sec. eq. 5	Aver.hydraulic conductivity, k 10^8, m/sec.	Ave.hydraulic conductivity, k 10^8, m/sec.
226.0-228.8	4.0		
228.8-251.4	0.68		
251.4-269.2	0.45		
269.2-283.3	0.42		
283.3-295.5	0.43		
226.0-295.5		0.66	0.81
291.0-301.7	0.43		
301.7-315.7	0.95		
315.7-358.8	0.12		
291.0-358.5		0.34	0.37
355.0-370.0	0.12		
370.0-391.9	0.040		
391.9-405.8	0.066		
405.8-422.5	0.022		
355.0-422.5		0.059	0.036
415.0-438.7	0.024		
438.7-448.7	0.039		
448.7-469.2	0.		
469.2-482.5	0.048		
415.0-482.5		0.024	0.022
410.0-438.7	0.024		
438.7-448.7	0.039		
448.7-469.2	0.		
469.2-477.5	0.048		
410.0-477.5		0.022	0.022
451.5-469.2	0.		
469.2-501.8	0.048		
451.5-501.8		0.031	0.048

Table III: Comparison of results obtained from the labelled slug test and the pumping tests. Well no. E1S, Mors, Denmark.

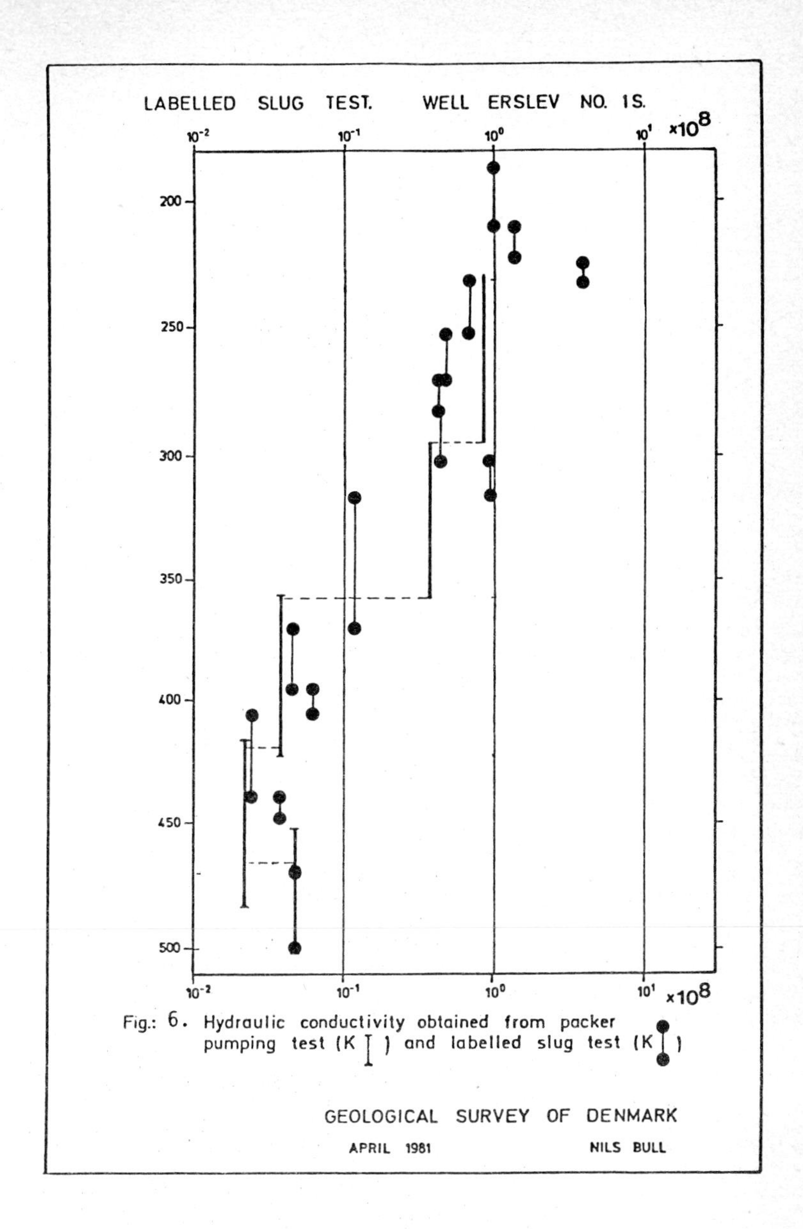

Fig.: 6. Hydraulic conductivity obtained from packer pumping test (K) and labelled slug test (K)

5. CONCLUSIONS

1. The hydraulic conductivity distribution calculated from the labelled slug test is in rather good agreement with the hydraulic conductivity distribution obtained from the results of the pumping tests.

2. A considerable improvement in the reslution of the permeability profile may be obtained by using LST.

3. LST is performed as a wire-line test, and no rig is required.

4. In relation to the amount of information obtained, the duration of the LST may be regarded as very short.

5. LST may be run prior to interval pumping or injection tests in order to localize high permeable zones. This may be important for the evaluation of the permeability of the fissures and may reduce the time necessary to obtain a representative water sample.

6. Leakage problems around the packers, associated with pumping and injection tests in low permeable formations, do not exist for the LST.

7. LST cannot totally eliminate the need for pumping and injection tests as no information about hydraulic conductivity, as a function of the distance from the bore hole, is provided by this method. Results of the LST could give misleading information if the bore-hole conditions have been influenced by the drilling process (plugging, invasion etc.).

6. REFERENCES

[1] Marine, I.V.: "Determination of the location and connectivity of fractures in metamorphic rock with in-hole tracers." Proc. of NEA/IAEA Workshop, Paris 19th-21st March, 1979. Low-flow, low-permeability, Paris, 1979.

[2] Gosk, E., N. Bull and L.J. Andersen: "Hydrogeological Programme, Mors Salt Dome." Hydrogeological Main Report. Geological Survey of Denmark, Copenhagen, April 1981. 192 pages and 98 figures. (Internal report.)

[3] Bear, J.: "Hydraulic of Groundwater", McGraw-Hill Series in Water Resources and Environmental Engineering, Jerusalem, 1979.

DISCUSSION

A.V. JOSHI, Denmark

Have you tried to correlate the results from the labelled slug test with permeabilities measured on cores ?

L.J. ANDERSEN, Denmark

As the method described has just been completed within the last 2 weeks, there has been no time to correlate the calculated values with other data than the pumping test data.

A.V. JOSHI, Denmark

Can you observe change in permeabilities where fissures are observed ?

L.J. ANDERSEN, Denmark

No.

A.V. JOSHI, Denmark

Can you distinguish the different types of chalk from results of the slug test ?

L.J. ANDERSEN, Denmark

This has not been attempted yet !

RECONNAISSANCE D'UN MASSIF ROCHEUX A GRANDE PROFONDEUR
MESURE DE L'ORIENTATION DES CAROTTES DE FORAGE
MESURE DE FAIBLES VALEURS DE PERMEABILITE

L. Bertrand, J.-P. Breton, B. Genetier, P. Vaubourg
Bureau de Recherches Géologiques et Minières
Orléans (France)

Résumé

L'étude de la fracturation comporte : l'orientation, la longueur, l'écartement et la répartition avec la profondeur des différentes fractures observées. L'appareillage mis en oeuvre est constitué d'un orienteur BTV 0-20 (brevet BRGM - CFFM) associé à un ensemble "Eastman multishot". Un programme de traitement informatique permet de repositionner les orientations en tenant compte des mesures de déviation du sondage.

Pour évaluer la perméabilité, on a retenu deux types d'essai : les "slug-tests" ou chocs hydrauliques et les injections entre obturateurs (packers). Les valeurs de perméabilité déduites des mesures sont de 3 à 7.10^{-11} m/s par la première méthode et de 3.10^{-11} à 3.10^{-9} m/s pour la deuxième. La mesure de perméabilités aussi faibles s'est révélée très délicate et a nécessité l'utilisation du matériel à ses limites de sensibilité. Les variations de perméabilité avec la profondeur étaient, sur le site testé, plus importantes que celles proposées par d'autres auteurs.

Abstract

To study fractured media, it is necessary to take into account several parameters such as : direction, length, spacing and distribution with depth of the different observed fractures. The instrumentation which has been used comprises a orientator BTV 0-20 (BRGM - CFFM patent) associated with a "Eastman multishot" equipment. A computer program enables to correct the directions, taking into account the deflection measurements in the borehole.

To estimate the hydraulic conductivity, two methods have been used : slug tests and packer tests. Following those measurements, the calculated permeabilities vary from 3 to 7.10^{-11} m/s with the first method and from 3.10^{-11} to 3.10^{-9} m/s with the second one. The measurement of such low permeabilities was very difficult and required to use the equipment at its sensitivity limitation. The permeability variations with depth appeared to be more important on the experimental site than results given by some other authors.

1. INTRODUCTION

Dans le cadre du programme de Recherche et Développement de la Commission des Communautés Européennes, l'effort de recherche principal a été consacré en France à l'analyse des possibilités offertes par les roches cristallines.

Après la reconnaissance en surface de divers massifs granitiques, le Commissariat à l'Energie Atomique, Institut de Protection et de Sûreté Nucléaire a décidé de faire exécuter un forage profond (ordre de 1 000 m) dans le massif granitique d'Auriat (30 km à l'Est de Limoges). Le C.E.A. en a confié l'exécution à l'une de ses filiales et a demandé au Bureau de Recherches Géologiques et Minières d'assurer la maîtrise d'oeuvre de ces travaux et de réaliser certaines études de reconnaissance à grande profondeur dont la mesure de l'orientation des carottes de forage et celle de la perméabilité.

2. RAPPEL DES RESULTATS DE L'ETUDE DE RECONNAISSANCE

Les principales conclusions de cette étude sont :

- le massif granitique (d'une superficie de 92 km^2) est nettement circonscrit dans son encaissant métamorphique ;

- le granite présente deux faciès :
  . l'un prépondérant à grain moyen porphyroïde,
  . l'autre à grain fin ;

- le granite a vraisemblablement une épaisseur minimale de 1 500 m ;

- la fracturation est importante à toutes les échelles, mais le granite porphyroïde est moins fracturé que le granite à grain fin ;

- entre 0 et 50 m de profondeur, le granite porphyroïde est moins perméable ($K = 6 \times 10^{-9}$ m/s) que le granite à grain fin ($K = 3 \times 10^{-8}$ m/s).

La zone où les forages ont été implantés a été sélectionnée en prenant pour critères :

- épaisseur maximale du granite (anomalie légère en gravimétrie),
- granite à grain moyen porphyroïde,
- granite peu fracturé en profondeur (électriquement résistant)
- possibilité d'accès, surface disponible, alimentation en eau, en électricité, etc...

Dans cette zone, il a été exécuté en carottage continu deux forages verticaux distants en surface de 10 m. Le forage principal (n° 689/2X/14) a atteint la profondeur de 1003,15 m, l'autre forage (n° 689/2X/13) a été arrêté à 504,40 m.

3. COUPES TECHNIQUES DES FORAGES

Le carottage en continu a été préféré au forage en destructif pour plusieurs raisons :

- la fracturation, les variations lithologiques du granite devaient pouvoir être observées dans les meilleures conditions ;

- l'orientation n'était possible qu'en carottage ;

- de nombreuses mesures en laboratoire étaient prévues. Il fallait pouvoir disposer de bons échantillons ;

- les vitesses d'avancement étaient plus grandes.

Les coupes techniques des deux forages sont représentées sur la figure 1. La déviation finale du forage principal (689/2X/14) est de 22 m par rapport à la verticale.

4. MESURE DE L'ORIENTATION DES CAROTTES DE FORAGE

Le granite constitue un milieu aquifère discontinu, dans lequel l'eau ne se déplace pour l'essentiel que suivant des fractures inégalement ouvertes et interconnectées. Il est donc important de connaître l'orientation et la fréquence des différentes familles de discontinuités recoupant un massif rocheux fracturé.

L'appareillage mis en oeuvre pour l'orientation est constitué d'un orienteur BTV 0-20 (brevet BRGM - CFFM) associé à un ensemble "Eastman multishot".

Quand une carotte sort d'un carottier, on connait sa cote et son sens (tête et pied) ; on peut connaître l'orientation de son grand axe qui est celle du sondage à la cote où elle a été prélevée, mais on ne connait pas sa position initiale autour de cet axe en raison des rotations indéterminées qu'elle a subies pendant le forage et au cours des manoeuvres. L'orienteur BTV détermine cette position initiale en prenant l'empreinte du fond de trou entre chaque passe et en repérant cette empreinte par rapport au Nord magnétique, ce qui permet de déterminer la position d'une génératrice des carottes de la passe suivante avant que celle-ci soit exécutée.

L'appareil se compose :

- d'un bloc portant une couronne de six palpeurs et un palpeur central

- d'un Eastman multishot utilisé ordinairement pour mesurer la déviation des forages. Il comprend une boussole, un clinomètre (unité d'angle) et une caméra dotée d'un mouvement d'horlogerie, qui prend une photographie de l'ensemble boussole-clinomètre toutes les minutes.

L'appareil a été utilisé à partir de 250 m, les carottes ont été orientées de 219 m à 1003 m dans le forage principal car on a pu reconstituer les carottes en continu de 250 m à 219 m et de 250 m à 400 m dans l'autre forage.

Parallèlement à ces mesures, les carottes ont fait l'objet d'un relevé systématique des paramètres structuraux :

. profondeur,
. type d'élément structural mesuré (diaclase, petite faille, grande faille, etc...),
. paramètres géométriques (orientation, ouverture, épaisseur, nature du remplissage, planéité, rugosité),
. paramètres cinématiques (stries, etc...).

L'ensemble de ces informations est ensuite traité au moyen d'un programme de calcul qui permet de repositionner toutes les fractures par rapport au Nord géographique, de tracer des diagrammes de densité (Schmidt).

Dans le forage principal, 3812 fractures ont ainsi été analysées ; elles ont pu être regroupées en sept familles principales. Ces familles de fractures se retrouvent tout au long du forage avec des orientations constantes ; seules les densités peuvent varier.

Au cours de l'étude préliminaire de surface du massif d'Auriat, seulement trois familles principales de fractures avaient été distinguées, la fracturation en profondeur parait donc plus complexe. Toutefois, il est probable que l'ensemble des mesures en surface regroupe

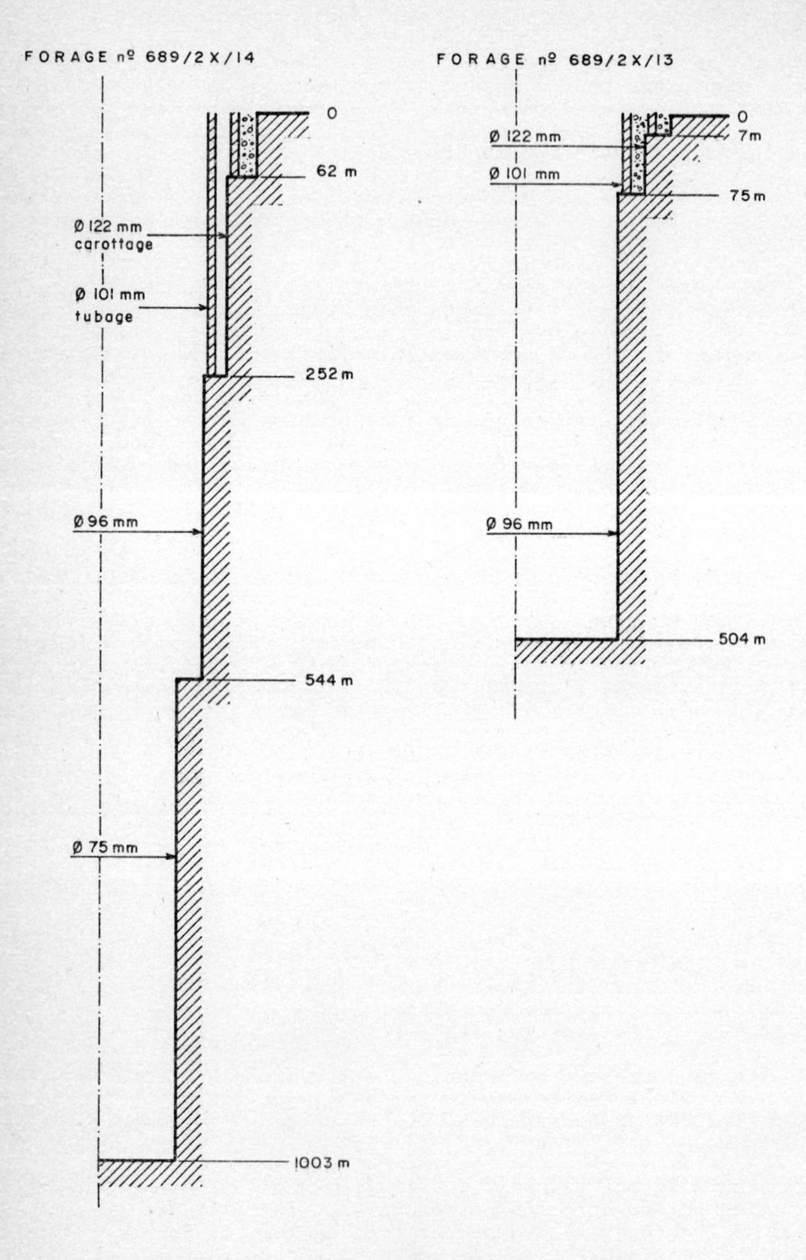

fig.1 - Coupes techniques des forages

les six familles subverticales (auxquelles s'ajoute la famille des fractures subhorizontales reconnues grâce au forage) autour de trois directions seulement, ceci en raison de variations assez faibles des orientations des familles principales de fracture.

5. MESURE DE FAIBLES VALEURS DE PERMEABILITE

Observation liminaire :

Le granite constitue un milieu aquifère discontinu, dans lequel l'eau ne se déplace pour l'essentiel que suivant des fissures inégalement ouvertes ou reliées entre elles. Décrire son comportement par des paramètres classiques de perméabilité et de transmissivité l'assimile en théorie à un milieu continu équivalent: la validité de cette équivalence est liée à la notion de volume élémentaire de référence, dont l'ordre de grandeur, pour le milieu granitique considéré, est au moins décamétrique. D'où la nécessité de "tester" le granite par des "passes" assez grandes pour être représentatives.

La notion de transmissivité implique de plus celle d'écoulement bidimensionnel dans un "monocouche", condition non rigoureusement réalisée dans un massif granitique : l'assimilation d'une hauteur de "tranche" explorée à une couche est naturellement une approximation, qui tend en général à surestimer la "perméabilité" déduite de la transmissivité apparente.

Les termes perméabilité et transmissivité seront donc employés ici dans les sens de perméabilité (ou conductivité hydraulique) et de transmissivité équivalentes.

Lors de l'étude de reconnaissance, la perméabilité moyenne sur les 50 premiers mètres avait été évaluée par pompage à 6.10^{-9} m/s. Il s'est confirmé très rapidement au cours des travaux de foration que le milieu granitique était certainement peu perméable ; en effet, on a noté :

- aucune perte de circulation décelable par les moyens de mesure du chantier,
- que les fractures fermées sont nombreuses,
- que l'altération hydrothermale est bien développée, ce qui laisse penser que les fractures identifiées comme ouvertes au sol sont très probablement colmatées,
- que les zones cataclasées sont recristallisées.

Aussi des pompages à débit constant n'étaient pas envisageables ; on a donc retenu deux types d'essai :

- les chocs hydrauliques ou "slug-tests"
- les injections entre obturateurs (packers).

5.1. Chocs hydrauliques ou slug-tests

L'expérience consiste à provoquer un abaissement instantané H_0 du niveau et à suivre la remontée de ce niveau en fonction du temps. Dans le cas d'un forage cylindrique, H. Cooper et al. [1] ont proposé une interprétation de ces essais par superposition des courbes expérimentales à des courbes types.

Dans le forage principal (n° 689/2X/14) trois slug-tests ont été réalisés à l'avancement et un une fois la foration terminée. L'abaissement du niveau d'eau a été obtenu soit par pompage, soit par soupapage à l'aide d'un train de tige équipé à son extrêmité inférieure d'un clapet ou d'un sabot (vidange par débordement). Dans tous les cas, le temps Δt nécessaire à l'exécution de l'opération de vidange est resté faible vis à vis de la durée d'observation de la remontée du niveau d'eau.

La superposition des données expérimentales sur les courbes types a aboutit aux résultats récapitulés dans le tableau I.

Tableau I

Résultats des interprétations des 3 premiers slug-tests dans le forage n° 689/2X/14

n° essai	Date	Trou nu de ... à ... m	e m	$\alpha = 10^{-4}$ T m^2/s	K m/s
1	12-14/04/80	61,80 à 251,95	190,15	$1,3 \times 10^{-8}$	7×10^{-11}
2	27/04-01/05/80	251,95 à 544,20	292,25	$0,9 \times 10^{-8}$	3×10^{-11}
3	21-27/05/80	544,20 à 1003,15	458,95	$0,7 \times 10^{-8}$	$1,5 \times 10^{-11}$

En supposant que tous les tubages isolent parfaitement les intervalles testés des niveaux supérieurs, on note que :

- le premier intervalle (61,80 m à 251,95 m), bien que le moins épais, est le plus transmissif,

- le troisième intervalle (544,20 m à 1003,15 m) a presque la même valeur de transmissivité que le deuxième qui est 0,6 fois moins épais.

Sachant que la perméabilité est égale au quotient de la transmissivité par l'épaisseur de l'aquifère, il y a donc une décroissance de la perméabilité avec la profondeur.

Pour le quatrième slug-test (trou nu de 251,95 à 1003,15 m), le niveau a été rabattu au moyen d'un train de tiges obturé à sa base. L'interprétation de cet essai est rendue plus complexe du fait de l'existence de nombreux changements de diamètre et du dénoyage très probable de certaines fractures productives. La remontée du niveau d'eau a été observée pendant 1600 h; par extrapolation des premières mesures, la profondeur initiale du niveau était de 615 m. La transmissivité calculée à partir de l'interprétation de ce quatrième test est de l'ordre de 1.10^{-8} m^2/s pour l'intervalle 251,95 à 1003,15 m ; pour ce même intervalle, la somme des transmissivités déduites des deuxième et troisième slug-tests est de $1,7.10^{-8}$ m^2/s, c'est-à-dire du même ordre de grandeur.

Dans l'autre forage (n° 689/2X/13), un slug-test a été réalisé dans le trou nu de 74,95 à 504,20 m ; la transmissivité calculée est de 1 à $1,5.10^{-8}$ m^2/s (perméabilité de 2,3 à $3,5.10^{-11}$ m/s). Elle est du même ordre de grandeur que les valeurs de transmissivité trouvées dans le forage principal (n° 689/2X/14) entre 61,80 et 544,20 m de profondeur.

5.2. Essais d'injection entre obturateurs

L'essai consiste à mesurer les débits d'eau injectés dans le terrain à des pressions croissantes et décroissantes maintenues constantes pendant la durée du palier. Dans le voisinage de la cavité d'essai (espace compris entre les deux obturateurs) de grande longueur par rapport à son diamètre, l'écoulement est approximativement radial plan, la perméabilité est calculée d'après l'équation des écoulements cylindriques en régime pseudo-permanent.

Compte tenu des faibles valeurs de perméabilité à mesurer, des précautions particulières ont été prises tant sur les conditions d'étanchéité des circuits que sur les dispositifs de mesure et d'enregistrement des débits et des pressions.

5.2.1. Description générale de l'équipement utilisé

La figure 2 montre les différents éléments du dispositif d'injection utilisé pour mesurer les perméabilités dans le forage principal (n° 689/2X/14). Les caractéristiques principales des appareils sont :

. Obturateurs ou packers :
Les essais sont effectués avec double obturateurs de marque TAM. On a choisi une hauteur d'injection de 10 m. Les obturateurs sont gonflés à l'eau simultanément à la pression de 150 bars; la longueur utile d'obturateur est de 1,20 m.

. Train de tiges :
L'étanchéité du train de tiges a été contrôlée à une pression de 120 bars grâce à un dispositif de clapet à billes éjectables situé entre le train de tiges et l'obturateur.

. Pompes d'injection :
L'injection d'eau sous pression peut être faite soit au moyen d'une pompe à piston (débit 0-60 l/mn, pression 0-120 bars), soit au moyen d'un surpresseur pneumatique (pression 400 bars) pour les débits très faibles.

. Dispositif de mesure et d'enregistrement :
La pression était mesurée au moyen de deux manomètres anti-vibratoires et d'un capteur électrique de pression à jauge de contrainte. Sur la conduite d'injection étaient montés en parallèle trois capteurs de débit instantané (0,18 m^3/h, 1 m^3/h et 5 m^3/h) avec totalisation des volumes écoulés. Un enregistreur analogique à 6 voies permet l'enregistrement simultané de la pression et du débit.

5.2.2. Déroulement des essais et résultats

Les essais ont été pratiqués :

- soit par injection entre doubles obturateurs distants de 10 m,
- ou par injection entre le fond du trou et un obturateur unique.

Dans le premier cas, les intervalles essayés ont été sélectionnés à l'aide :

- des logs de fracturation (indice de fracturation et R.Q.D. (Rock Quality Designation)),
- d'indications pétrographiques (zones cataclasées par exemple),
- des diagraphies, essentiellement résistivité et sonique; en particulier toutes les anomalies soniques ont été retenues.

Entre 251 et 544 m, quatorze intervalles ont été retenus. On s'est assuré par un examen visuel des carottes que les obturateurs

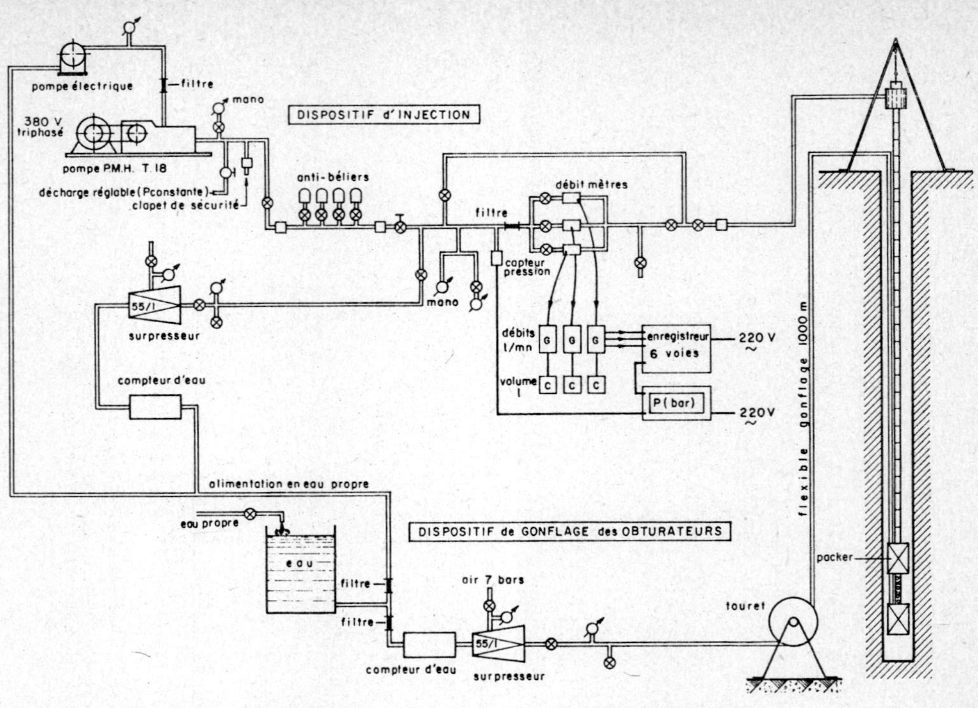

Fig. 2 - Schéma du dispositif de mesure de perméabilités

seraient ancrés dans des zones compactes où le risque de contournement de l'obturateur pendant les injections serait minime. Le contrôle permanent du niveau d'eau en surface a permis de vérifier qu'il n'y avait pas de contournement des obturateurs pendant les injections.

5.2.3. Résultats

Les résultats des essais d'injection entre deux obturateurs distants de 10 m sont récapitulés dans le tableau II :

TABLEAU II

Résultats de l'interprétation des essais d'injection entre deux obturateurs dans le forage n° 0689/2X/14

N° de l'intervalle	Intervalle testé de ... à ...	T m^2/s	K m/s
1	520 - 530 m	$< 0,22 \times 10^{-9}$	$< 2,2 \times 10^{-11}$
2	495 - 505 m	$0,2 \times 10^{-9}$	$2,2 \times 10^{-11}$
3	460 - 470 m	$0,7 \times 10^{-9}$	$7,2 \times 10^{-11}$
4	440 - 450 m	$2,1 \times 10^{-9}$	$2,1 \times 10^{-10}$
5	400 - 410 m	$3,0 \times 10^{-9}$	$3,0 \times 10^{-10}$
6	390 - 400 m	$< 0,2 \times 10^{-9}$	$< 2,0 \times 10^{-11}$
7	376 - 386 m	$< 6,5 \times 10^{-9}$	$< 6,5 \times 10^{-10}$
8	365 - 375 m	$9,5 \times 10^{-9}$	$9,5 \times 10^{-10}$
9	330 - 340 m	$< 0,3 \times 10^{-9}$	$3,0 \times 10^{-11}$
10	320 - 330 m	$< 0,15 \times 10^{-9}$	$1,5 \times 10^{-11}$
11	310 - 320 m	$< 0,25 \times 10^{-9}$	$< 2,5 \times 10^{-11}$
12	298 - 308 m	30×10^{-9}	$3,0 \times 10^{-9}$
13	287 - 297 m	10×10^{-9} (?)	$1,0 \times 10^{-9}$ (?)
14	270 - 280 m	$3,0 \times 10^{-9}$	$3,0 \times 10^{-10}$

Deux intervalles - 298-308 et 365-375 m - apparaissent comme beaucoup plus perméables. Leurs perméabilités sont de 10 à plus de 100 fois supérieures à celles des autres intervalles. L'étude structurale a montré que les failles normales - failles ayant joué en distension - étaient peu nombreuses. Il en existe précisément à la cote 301,80 m.

Deux essais d'injection avec obturateur unique ont été réalisés, l'un. Pour l'un, l'obturateur était posé à 256 m et pour l'autre à 530,40 m. Les résultats obtenus sont rassemblés dans le tableau III.

TABLEAU III

Résultats de l'interprétation des essais d'injection avec obturateur unique dans le forage n° 0689/2X/14

Intervalle testé	Hauteur testée e (m)	T m^2/s	K m/s
530,40 à 1003,15 m	472,75	$< 4,2 \times 10^{-10}$	$< 1 \times 10^{-12}$
256,00 à 1003,15 m	747,15	$7,4 \times 10^{-8}$	$\sim 1 \times 10^{-10}$

5.3. Conclusions

Les tests qui ont été choisis sont bien adaptés au type de milieu que nous avons rencontré.

Les tests hydrauliques ont permis de mettre en évidence une diminution des valeurs de la perméabilité moyenne du granite avec la profondeur :

- de 0 à 50 m (étude de reconnaissance) — $K = 6 \times 10^{-9}$ m/s (pompage d'essai)
- de 60 à 250 m environ — $K = 7 \times 10^{-11}$ m/s (slug test)
- de 250 à 540 m environ — $K = 3,4 \times 10^{-11}$ m/s (slug test) ou $2,7 \times 10^{-10}$ m/s (injection)
- de 540 à 1003,15 m environ — $K < 1.10^{-12}$ m/s (injection)

On voit qu'au delà de 500 m de profondeur les valeurs de la perméabilité sont extrêmement faibles et qu'elles s'apparentent plus aux valeurs que l'on attribue habituellement à la matrice du granite.

Des perméabilités dix à cent fois supérieures, associées à des niveaux fracturés très localisés, ont cependant été trouvées à :

- 298 - 308 m — $K = 3 \times 10^{-9}$ m/s
- 365 - 375 m — $K = 9,5 \times 10^{-10}$ m/s.

La perméabilité moyenne mesurée par slug-test entre 74,95 m et 504,4 m dans le forage n° 0689/2X/13 est du même ordre de grandeur que celle obtenue sur le forage n° 0689/2X/14 par la même méthode, soit 3 à $3,4 \times 10^{-11}$ m/s.

Le tableau IV ci-dessous permet la comparaison des résultats acquis dans le granite d'Auriat avec ceux établis dans d'autres massifs granitiques aussi bien en France qu'à l'étranger.

Enfin, sur la figure n° 3, on pourra comparer la distribution des perméabilités moyennes en profondeur avec celle qui est proposée par différents auteurs.

TABLEAU IV

Quelques valeurs de perméabilité
en milieu granitique (France et étranger)

Lieu	Nature du terrain *	Méthode	Profondeur	K m/s	Réf. n°
EN FRANCE					
. Massif "A"	monzogranite porphyroïde	pompage	0 - 40 m	1,5 à 3 . 10^{-6}	[2]
. Massif "B"	granodiorite à biotite seule	pompage	50 - 100 m	1,6 . 10^{-7}	[3]
. Massif d'Auriat	monzogranite porphyroïde	pompage	0 - 50 m	6 . 10^{-9}	[4]
A L'ETRANGER					
. Carmellis (Grande-Bretagne)	granite à grain grossier	slug-test	25 - 286 m	1,2 . 10^{-9}	[5]
. Cornouailles (Grande-Bretagne)		packer-test	25 - 286 m	1,1 . 10^{-9}	
. Suède	granite à grain moyen	packer-test	1 - 140 m	1 à 4 . 10^{-8}	[6]
		drainage	-	4 . 10^{-8}	
. Stripamine (Suède)		packer-test	471 - 881 m	2,2 . 10^{-9}	[7]
			471 - 511 m	> 10^{-10} max 2,6 . 10^{-8}	
			511 - 781 m	très très faible	
			781 - 881 m	10^{-9} < K < 10^{-11}	[8]
		drainage	300 m	3 à 9 . 10^{-11}	

* faciès dominant

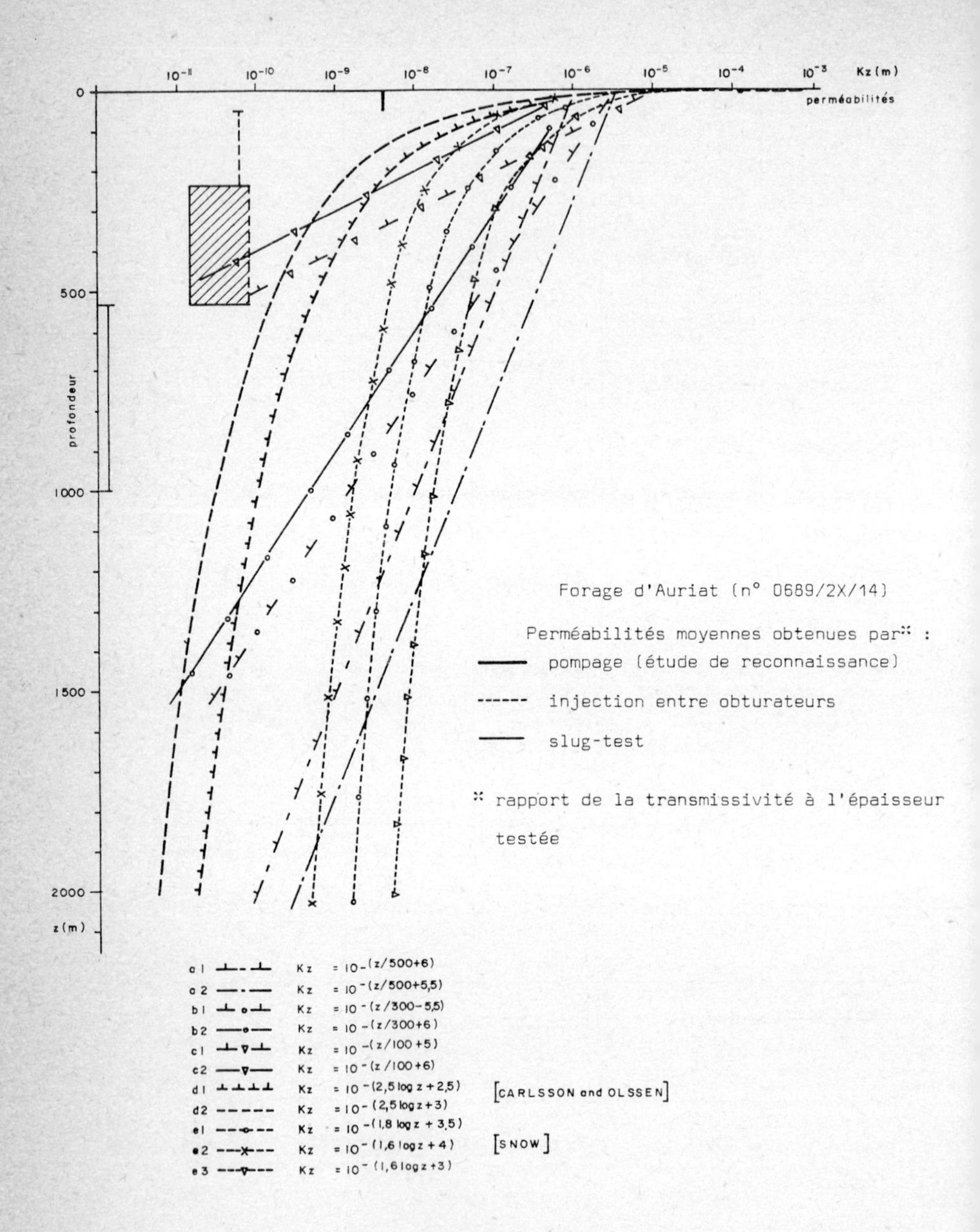

Fig. 3 - Lois de variation des perméabilités en fonction de la profondeur

BIBLIOGRAPHIE

[1] COOPER (H.H.), BREDEHOEFT (J.D.), PAPADOPOULOS (I.S.) 1973 .- On the analysis of "slug-test" data .- Water Resources Research, vol. 9, n° 4, p. 1087-1089.

[2] Anonyme (1978) .- Modalités de transfert de produits radioactifs dans l'environnement géologique. Essais in situ sur le massif "A". - Rapport B.R.G.M. n° 78 SGN 166 HYD.

[3] Anonyme (1978) .- Massif "B" - Reconnaissance du milieu et mesures in situ des paramètres. - Rapport B.R.G.M.

[4] Anonyme (1980). - Etude préliminaire du massif "K" .- Rapport B.R.G.M. n° 80 SGN 129 HYD.

[5] BLACK (J.H.) 1979 .- Results of a multiple borehole pumping test in low permeability granite .- Proceedings of the NEA/IAEA Workshop, OECD Nuclear Energy, Paris 1979, p. 183-195.

[6] LINDBLOM (U.E.) 1979 .- Comparison of low and permeability interpreted from in situ measurements in a granitic rock .- Proceedings of the NEA/IAEA Workshop, OECD Nuclear Energy, Paris 1979, p. 125 -136.

[7] CARLSSON (L.) 1979 .- Estimation of hydraulic conductivity in Swedish Precambrien Crystalline Bedrock .- Proceedings of the NEA /IAEA Workshop, OECD Nuclear Energy, Paris 1979, p. 97-113.

[8] WITHERSPOON (P.A.) 1980 .- Large scale permeability measurements in fractured crystalline rock .- International Geologic Congress, Paris.

SESSION V

Chairmen - Présidents

F. GERA

(Italy)

J.D. MATHER

(United Kingdom)

SEANCE V

GENERAL DISCUSSION AND CONCLUSIONS

DISCUSSIONS GENERALES ET CONCLUSIONS

Rapporteurs :

P.B. GREENWOOD (United Kingdom)
N. VANDENBERGHE (Belgium)
B. NEERDAEL (Belgium)

INTRODUCTION

This report summarises the discussions held during the working sessions. The topics are covered in the order in which they were discussed. General conclusions are outlined and differences in views noted. The basis for these discussions was the list of questions drawn up after the formal sessions in which papers were presented and the work of the various participants was reviewed.

DISCUSSION ITEMS

- Value of general criteria.
- Required criteria in different phases of siting.
- Relationships between siting criteria and radionuclide migration to the biosphere and assessment of consequences.
- Host-rock dependent criteria.
- Ranking of siting criteria.
- Preliminary ranking of sites.
- Emplacement concepts : deep hole vs. conventional mine-investigation methods for siting.
- Impact of waste retrievability on siting.
- Permeability determination.
- Technical performance of testing methods.
- Should areas be reserved for radioactive waste disposal and how should they be chosen ?
- Should risk of future human intrusion be an important consideration ?
- Public acceptance and/or information.

SUMMARY OF DISCUSSIONS

Value of General Criteria

In the USA, a systematic survey of areas for the siting of repositories is underway which is similar to, but more detailed than, that used in CEC countries* for the production of a catalogue of geological waste repositories. But general criteria and statements of earth science data requirements are difficult to formulate because of the wide variations in host rocks available, their geological setting and the waste types to be accommodated. Lists of general criteria and data needs to encompass all situations will be long and both the quantification and ranking of such criteria dependent on geological factors, interrelated quantities, trade-offs and the nature of engineered barriers to be used.

The presentation and interpretation of general criteria could give rise to misconceptions and the participants concluded that the words "guidelines" and "factors" are to be preferred to "criteria" and "requirements" in this context. The only general statement about these that can be made concerns the necessity of determining those required to establish the safety case for repository operation and performance.

Required Criteria in Different Phases of Siting

The participants recognised that the criteria would be similar for different phases of the siting process, but that the data requirements would be different. The factors to be considered will depend on the degree to which a potential site has been localised, the data already available and their influence on safety. There is sufficient flexibility for different countries to adopt different approaches. For example, different countries may use different types of engineered barrier to supplement the natural barriers available at potential disposal sites. A paper on criteria is being prepared by the IAEA.

Host-Rock Dependent Criteria

Two types of situation for repositories were recognised : those localised to a rather small area (example, salt domes) and those which might be placed within widely distributed rock (crystalline, argillaceous or bedded salt). In the former case and in small countries the utilisation of salt domes for other purposes must be considered. A major problem in the assessment of the influence of the alternative use of a potential site is that values have to be estimated and may change in the future. The importance to be attached to tectonic stability will be dependent on whether the host rock is plastic or brittle.

Ranking of Sites

The participants heard a description of the process by which salt domes on the gulf coast in the USA had been evaluated. Broad criteria such as size and depth had been used to reduce some 500 domes considered to 8 candidates. Of these 8, 3 were eliminated because they were considered to be of insufficient extent or depth, and one because of its former use ; but the 4 remaining were considered against many criteria including geological, environmental and socio-economic factors. On these grounds, the dome most favoured would be investigated further ; if it was finally eliminated, attention would pass to the next. No quantification or weighting of the criteria had been attempted yet.

---

* European catalogue of geological formations having suitable characteristics for the disposal of solidified high-level and/or long-lived radioactive waste. EUR 6891E (1981), (to be published).

The participants learned of the procedure by which different European countries had focused their attention on different areas to be used for research purposes. Most countries reported that ranking was being avoided ; an overview of the qualities and characteristics of different rock types and situations was being sought. To fix research sites, very general considerations have been taken into account, including, for example, geological, administrative and land ownership factors. Only the US seems to have reached the point where it is necessary to rank alternative sites. It was recognised that the options available to some countries are rather limited and the problem can be largely avoided. In many cases it was considered that the information available on potential areas was insufficient for them to be ranked at all.

Emplacement Concepts : Deep Holes vs. Mined Repositories

The participants recognised that the cost of the disposal of heat generating waste may depend on the emplacement method and quantity of waste involved ; in Denmark, it has been estimated that, for the small volume of waste foreseen by the Danish nuclear programme, the cost of deep borehole emplacement would be approximately one-tenth of disposal in a mined repository. The advisability of adopting deep borehole emplacement depends on the technical ability to determine rock properties around potential disposal boreholes. While some remote sensing methods are available and improvements can be expected, it was recognised that at present fractures, etc. may pass undetected. This may not be important in all situations and the method can be considered for some argillaceous and evaporatic rocks when retrievability is not considered necessary. It was recognised that at the present state of the development of remote sensing techniques, the mined repository concept gives more information about the geological conditions. It was also recognised that, for stratified and/or other predictable formations, the deep hole concept could be considered to be more applicable.

The difficulty of the deep hole concept related to the uncertainty about the geology surrounding the drill hole was illustrated by examples of granitic rocks at great depth which showed non-negligible permeability. Examples were drawn from well information, tunnelling and mining experiences in different countries. It was stressed that the nature of fractures (open or filled with alteration products, silica cement) is finally determining the permeability of the system rather than fracturing degree alone.

Impact of Waste Retrievability on Siting

From a technical viewpoint, there seems to be no need for the retrievability of the waste disposed. Indeed, it could even be argued that it would be better if the repository was sealed off quickly. Retrievability times of 50 or 100 years, as sometimes proposed, are anyhow arbitrary time periods. In case retrievability is for some reason needed, this requirement can have an implication for the site selection and the repository concept as well.

Permeability Determination and Movement of Radionuclides

As ground water is certainly a key factor in considering possible mobilisation of radionuclides, problems of measuring the hydraulic conductivity in low permeable media were considered separately. Especially, the difficulty was noted of obtaining increasingly precise measurements in rocks of decreasing permeability. It was felt that current theory and terminology for fissured media were not entirely adequate. While determination of the effective porosity distribution in larger fracture systems is achievable by field tests, in small fracture systems it still seems very difficult to work out the effective porosity distribution along the flow path. In this connection, it was stated that the relatively long times involved could present a limitation to the use of field experiments and laboratory tests could be considered.

As in fact not only the water flow itself but rather the movement of the radionuclides is the factor searched for, experiments are carried out to determine retardation factors under forced flow conditions. Diffusion of radionuclides from the flowing ground water into rock mass is an important phenomena even into a granitic rock. Information about radionuclide diffusion in rock masses either from nuclear explosion tests, uranium mining or the OKLO studies seems to be not particularly useful for the waste disposal problem because the available information is not detailed enough or is not related to the waste disposal environment. However, geochemical data in connection to particular mineral deposits and other geological conditions similar to those that might exist at disposal sites can provide useful information on possible migration of radionuclides.

Technical Performance of Testing Methods

It would be useful for hydrology researchers in the field of low permeability rocks to exchange information related to the availability and the performance of the particular pumping and monitoring equipment needed in this field. In the case of small yields of pumping tests, one should be careful to consider all influencing factors.

It was suggested that in long duration tests a computer system, continuously recording the data becoming available, could avoid the loss of data previous to an unforeseen or accidental interruption of the test.

Labelled slug test was suggested as a technique to obtain information about the vertical distribution of the hydraulic conductivity which is important in choosing of the intervals to be tested further.

The optimum number of boreholes required for the reconnaissance of a particular site was discussed. It was felt that geophysical data at the site should aid in forming the base for a sound judgement of the required number of boreholes. It was mentioned that, in the USA, a computer modelling technique is in preparation, with the potential to optimise the number of boreholes required to maximise the information from the data obtained.

Reserved Areas - Risks of Future Human Intrusion

The use of underground space has been growing rapidly in the past years and even new applications are foreseeable in the future. Mining activities of raw materials can be economically justified very near possible repository sites. As both the use of underground space and mining raw materials could eventually be competitors with a radioactive waste repository for a specific site, a long-term planning should be considered by the political authorities. The necessary legal measures should be taken as well. Obviously, if the need for deep raw materials and the use of underground space should increase and result in extensive exploratory or production drilling, this would considerably increase the risk of human intrusion into a repository. In this connection the case of shallow salt domes was mentioned, as being attractive sites for human activities.

The risk of future human intrusion is related to many factors including the question of future institutional control of disposal sites. In the case of scenarios where institutional control is considered to be lost relatively early after disposal the risk of human intrusion could become an important siting factor.

Public Acceptance and/or Information

It was the general feeling of the participants that waste repository siting is subject to the rules of democratic decision-making and therefore the handling of public acceptance falls into the political domain.

It is part of the responsibility of scientists and technicians to make the information on the subject available, in a suitable form, to decision-makers and members of the public requiring it.

LIST OF PARTICIPANTS LISTE DES PARTICIPANTS

BELGIUM - BELGIQUE

NEERDAEL, B., Centre d'Etude de l'Energie Nucléaire, CEN/SCK, Boeretang 200, B-2400 Mol

VANDENBERGHE, N., Belgian Geological Survey, Jennerstraat 13, B-1040 Brussels

DENMARK - DANEMARK

ANDERSEN, J.L., Geological Survey of Denmark, 31 Thoravej, DK-2400 Copenhagen NV

GOSK, E., Geological Survey of Denmark, 31 Thoravej, DK-2400 Copenhagen NV

JENSEN, H., The Inspectorate of Nuclear Installations, P.O. Box 217, DK-4000 Roskilde

JOSHI, A.V., Elsam, DK-7000 Fredericia

LETH, K., National Agency of Environmental Protection, 29 Strandgade, DK-1401 Copenhagen K

L. MORTENSEN, L., National Agency of Environmental Protection, Strandgade 29, DK-1401 Copenhagen K

FINLAND - FINLANDE

HÄRKÖNEN, H., Teollisuuden Voima Oy, TVO Power Company, Kutojantie 8, SF-02630 Espoo 63

JAKOBSSON, K.O., Institute of Radiation Protection, P.O. Box 268, SF-00101 Helsinki 10

PELTONEN, E.K., Technical Research Centre of Finland, Nuclear Engineering Laboratory, P.O. Box 169, SF-00181 Helsinki 18

FRANCE

BARBREAU, A., Commissariat à l'Energie Atomique, Institut de Protection et de Sûreté Nucléaire, CSDR, B.P. n° 6, F-92260 Fontenay-aux-Roses

BARTHOUX, A., Commissariat à l'Energie Atomique, Agence Nationale pour la Gestion des Déchets Radioactifs (ANDRA), 31-33 rue de la Fédération, F-75015 Paris

GENETIER, B., Bureau de Recherches Géologiques et Minières, Département Eau, B.P. 6009, F-45060 Orléans Cedex

GOBLET, P., Chercheur au Centre d'Informatique Géologique, Ecole des Mines de Paris, 35 rue Saint-Honoré, F-77305 Fontainebleau Cedex

de MARSILY, G., Directeur du Centre d'Informatique Géologique, Ecole des Mines de Paris, 35 rue Saint-Honoré, F-77305 Fontainebleau Cedex

MASURE, P., Bureau de Recherches Géologiques et Minières, B.P. 6009, F-45060 Orléans Cedex

ITALY - ITALIE

BRONDI, A., Comitato Nazionale per l'Energia Nucleare, Divisione Protezione Ambiente, Laboratorio Geologia Ambientale, CSN-Casaccia, CP 2400, I-00100 Roma

GERA, F., ISMES, Via T. Taramelli 14, I-00197 Roma

NORWAY - NORVEGE

HUSEBY, S., Geological Survey of Norway, Oslo-Kontoret, Drammensveien 230, Oslo 2

SWEDEN - SUEDE

AHLBOM, K., Geological Survey of Sweden, Box 670, S-751 28 Uppsala

HULT, A.E., KBS, Box 5864, S-102 48 Stockholm

SCHERMAN, S., National Council for Radioactive Waste, Box 5864, S-102 48 Stockholm

SWITZERLAND - SUISSE

THURY, M.F., NAGRA, Parkstrasse 23, CH-5401 Baden

UNITED KINGDOM - ROYAUME-UNI

BOURKE, P.J., Atomic Energy Research Establishment Harwell, Chemical Technical Division, B 151, AERE, Harwell, Oxon OX11 ORA

BROWNING, R., Department of the Environment, Room 307, Becket House, 1 Lambeth Palace Road, London SE1

GREENWOOD, P.B., Institute of Geological Sciences, Bld. 151, Harwell Laboratory, Didcot, Oxon

MATHER, J.D., Institute of Geological Sciences, Bld. 151, Harwell Laboratory, Didcot, Oxon

WILKS, R.S., Manager, Waste Management Section, Research and Development Department, British Nuclear Fuels Limited, Room J615, Risley, Warrington WA3 6AS

UNITED STATES - ETATS-UNIS

BEDINGER, M.S., US Geological Survey, Denver Federal Center, Box 25046, Mail Stop 417, Lakewood, Colorado 80225

SMITH, S.S., Project Manager, Office of NWTS Integration, Site Program Office, Office of Nuclear Waste Isolation, Battelle Memorial Institute, 505 King Avenue, Columbus, Ohio 43201

SWANSON, O.E., Site Programme Office, Office of Nuclear Waste Isolation, Battelle Memorial Institute, 505 King Avenue, Columbus, Ohio 43201

COMMISSION OF THE EUROPEAN COMMUNITIES
COMMISSION DES COMMUNAUTES EUROPEENNES

HAIJTINK, B., Commission of the European Communities, DG Research Science and Education, 200 rue de la Loi, B-1049 Bruxelles (Belgium)

INTERNATIONAL ATOMIC ENERGY AGENCY
AGENCE INTERNATIONALE DE L'ENERGIE ATOMIQUE

DLOUHY, Z., Division of Nuclear Safety and Environment Protection, International Atomic Energy Agency, B.P. Box 100, A-1400 Vienna (Austria)

SECRETARIAT

RÜEGGER, B., Division of Radiation Protection and Waste Management, Nuclear Energy Agency, 38 boulevard Suchet, F-75016 Paris (France)

SOME NEW PUBLICATIONS OF NEA

QUELQUES NOUVELLES PUBLICATIONS DE L'AEN

ACTIVITY REPORTS

Activity Reports of the OECD Nuclear Energy Agency (NEA)

- 8th Activity Report (1979)
- 9th Activity Report (1980)

RAPPORTS D'ACTIVITÉ

Rapports d'activité de l'Agence de l'OCDE pour l'Énergie Nucléaire (AEN)

- 8e Rapport d'Activité (1979)
- 9e Rapport d'Activité (1980)

Free on request – Gratuits sur demande

Annual Reports of the OECD HALDEN Reactor Project

- 19th Annual Report (1978)
- 20th Annual Report (1979)

Rapports annuels du Projet OCDE de réacteur de HALDEN

- 19e Rapport annuel (1978)
- 20e Rapport annuel (1979)

Free on request – Gratuits sur demande

• • •

INFORMATION BROCHURES

- The NEA Data Bank
- International Co-operation for Safe Nuclear Power
- NEA at a Glance
- OECD Nuclear Energy Agency: Functions and Main Activities

BROCHURES D'INFORMATION

- La Banque de Données de l'AEN
- Une coopération internationale pour une énergie nucléaire sûre
- Coup d'œil sur l'AEN
- Agence de l'OCDE pour l'Énergie Nucléaire : Rôle et principales activités

Free on request – Gratuits sur demande

• • •

SCIENTIFIC AND TECHNICAL PUBLICATIONS

PUBLICATIONS SCIENTIFIQUES ET TECHNIQUES

NUCLEAR FUEL CYCLE

LE CYCLE DU COMBUSTIBLE NUCLÉAIRE

Nuclear Fuel Cycle Requirements and Supply Considerations, Through the Long-Term (1978)

Besoins liés au cycle du combustible nucléaire et considérations sur l'approvisionnement à long terme (1978)

£4.30 US$8.75 F35,00

World Uranium Potential — An International Evaluation (1978)

Potentiel mondial en uranium — Une évaluation internationale (1978)

£7.80 US$16.00 F64.00

Uranium — Resources, Production and Demand (1979)

Uranium — ressources, production et demande (1979)

£8.70 US$19.50 F78,00

• • •

RADIATION PROTECTION

RADIOPROTECTION

Iodine-129
(Proceedings of an NEA Specialist Meeting, Paris, 1977)

Iode-129
(Compte rendu d'une réunion de spécialistes de l'AEN, Paris, 1977)

£3.40 US$7.00 F28,00

Recommendations for Ionization Chamber Smoke Detectors in Implementation of Radiation Protection Standards (1977)

Recommandations relatives aux détecteurs de fumée à chambre d'ionisation en application des normes de radioprotection (1977)

Free on request — Gratuit sur demande

Radon Monitoring
(Proceedings of the NEA Specialist Meeting, Paris, 1978)

Surveillance du radon
(Compte rendu d'une réunion de spécialistes de l'AEN, Paris, 1978)

£8.00 US$16.50 F66,00

Management, Stabilisation and Environmental Impact of Uranium Mill Tailings
(Proceedings of the Albuquerque Seminar, United States, 1978)

Gestion, stabilisation et incidence sur l'environnement des résidus de traitement de l'uranium
(Compte rendu du Séminaire d'Albuquerque, États-Unis, 1978)

£9.80 US$20.00 F80,00

Exposure to Radiation from the Natural Radioactivity in Building Materials
(Report by an NEA Group of Experts, 1979)

Exposition aux rayonnements due à la radioactivité naturelle des matériaux de construction
(Rapport établi par un Groupe d'experts de l'AEN, 1979)

Free on request — Gratuit sur demande

Marine Radioecology
(Proceedings of the Tokyo Seminar, 1979)

Radioécologie marine
(Compte rendu du Colloque de Tokyo, 1979)

£9.60 US$21.50 F86.00

Radiological Significance and Management of Tritium, Carbon-14, Krypton-85 and Iodine-129 arising from the Nuclear Fuel Cycle (Report by an NEA Group of Experts, 1980)

Importance radiologique et gestion des radionucléides : tritium, carbone-14, krypton-85 et iode-129, produits au cours du cycle du combustible nucléaire (Rapport établi par un Groupe d'experts de l'AEN, 1980)

£8.40 US$19.00 F76,00

The Environmental and Biological Behaviour of Plutonium and Some Other Transuranium Elements (Report by an NEA Group of Experts, 1981) (in preparation)

Le comportement mésologique et biologique du plutonium et de certains autres éléments transuraniens (Rapport établi par un Groupe d'experts de l'AEN, 1981) (en préparation)

£ US$ F

• • •

RADIOACTIVE WASTE MANAGEMENT

GESTION DES DÉCHETS RADIOACTIFS

Objectives, Concepts and Strategies for the Management of Radioactive Waste Arising from Nuclear Power Programmes (Report by an NEA Group of Experts, 1977)

Objectifs, concepts et stratégies en matière de gestion des déchets radioactifs résultant des programmes nucléaires de puissance (Rapport établi par un Groupe d'experts de l'AEN, 1977)

£8.50 US$17.50 F70,00

Treatment, Conditioning and Storage of Solid Alpha-Bearing Waste and Cladding Hulls (Proceedings of the NEA/IAEA Technical Seminar, Paris, 1977)

Traitement, conditionnement et stockage des déchets solides alpha et des coques de dégainage (Compte rendu du Séminaire technique AEN/AIEA, Paris, 1977)

£7.30 US$15.00 F60,00

Storage of Spent Fuel Elements (Proceedings of the Madrid Seminar, 1978)

Stockage des éléments combustibles irradiés (Compte rendu du Séminaire de Madrid, 1978)

£7.30 US$15.00 F60,00

In Situ Heating Experiments in Geological Formations (Proceedings of the Ludvika Seminar, Sweden, 1978)

Expériences de dégagement de chaleur in situ dans les formations géologiques (Compte rendu du Séminaire de Ludvika, Suède, 1978)

£8.00 US$16.50 F66,00

Migration of Long-lived Radionuclides in the Geosphere (Proceedings of the Brussels Workshop, 1979)

Migration des radionucléides à vie longue dans la géosphère (Compte rendu de la réunion de travail de Bruxelles, 1979)

£8.30 US$17.00 F68,00

Low-Flow, Low-Permeability Measurements in Largely Impermeable Rocks (Proceedings of the Paris Workshop, 1979)

Mesures des faibles écoulements et des faibles perméabilités dans des roches relativement imperméables (Compte rendu de la réunion de travail de Paris, 1979)

£7.80 US$16.00 F64,00

On-Site Management of Power Reactor Wastes (Proceedings of the Zurich Symposium, 1979)

Gestion des déchets en provenance des réacteurs de puissance sur le site de la centrale (Compte rendu du Colloque de Zurich, 1979)

£11.00 US$22.50 F90,00

Recommended Operational Procedures for Sea Dumping of Radioactive Waste (1979)

Recommandations relatives aux procédures d'exécution des opérations d'immersion de déchets radioactifs en mer (1979)

Free on request — Gratuit sur demande

Guidelines for Sea Dumping Packages of Radioactive Waste (Revised version, 1979)

Guide relatif aux conteneurs de déchets radioactifs destinés au rejet en mer (Version révisée, 1979)

Free on request — Gratuit sur demande

Use of Argillaceous Materials for the Isolation of Radioactive Waste (Proceedings of the Paris Workshop, 1979)

Utilisation des matériaux argileux pour l'isolement des déchets radioactifs (Compte rendu de la Réunion de travail de Paris, 1979)

£7.60 US$17.00 F68,00

Review of the Continued Suitability of the Dumping Site for Radioactive Waste in the North-East Atlantic (1980)

Réévaluation de la validité du site d'immersion de déchets radioactifs dans la région nord-est de l'Atlantique (1980)

Free on request — Gratuit sur demande

Decommissioning Requirements in the Design of Nuclear Facilities (Proceedings of the NEA Specialist Meeting, Paris, 1980)

Déclassement des installations nucléaires : exigences à prendre en compte au stade de la conception (Compte rendu d'une réunion de spécialistes de l'AEN, Paris, 1980)

£7.80 US$17.50 F70,00

Borehole and Shaft Plugging (Proceedings of the Columbus Workshop, United States, 1980)

Colmatage des forages et des puits (Compte rendu de la réunion de travail de Columbus, États-Unis, 1980)

£12.00 US$30.00 F120,00

Radionucleide Release Scenarios for Geologic Repositories (Proceedings of the Paris Workshop, 1980)

Scénarios de libération des radionucléides à partir de dépôts situés dans les formations géologiques (Compte rendu de la réunion de travail de Paris, 1980)

£6.00 US$15.00 F60,00

Cutting Techniques as related to Decommissioning of Nuclear Facilities (Report by an NEA Group of Experts, 1981)

Techniques de découpe utilisées au cours du déclassement d'installations nucléaires (Rapport établi par un Groupe d'experts de l'AEN, 1981)

£ 3.00 US$ 7.50 F 30.00

Decontamination Methods as related to Decommissioning of Nuclear Facilities (Report by an NEA Group of Experts, 1981)

Méthodes de décontamination relatives au déclassement des installations nucléaires (Rapport établi par un Groupe d'experts de l'AEN, 1981)

£ 2.80 US$ 7.00 F 28,00

• • •

SAFETY	SÛRETÉ
Safety of Nuclear Ships (Proceedings of the Hamburg Symposium, 1977)	Sûreté des navires nucléaires (Compte rendu du Symposium de Hambourg, 1977)

£17.00 US$35.00 F140,00

Nuclear Aerosols in Reactor Safety (A State-of-the-Art Report by a Group of Experts, 1979)	Les aérosols nucléaires dans la sûreté des réacteurs (Rapport sur l'état des connaissances établi par un Groupe d'Experts, 1979)

£8.30 US$18.75 F75,00

Plate Inspection Programme (Report from the Plate Inspection Steering Committee – PISC – on the Ultrasonic Examination of Three Test Plates), 1980	Programme d'inspection des tôles (Rapport du Comité de Direction sur l'inspection des tôles – PISC – sur l'examen par ultrasons de trois tôles d'essai au moyen de la procédure «PISC» basée sur le code ASME XI), 1980

£3.30 US$7.50 F30.00

Reference Seismic Ground Motions in Nuclear Safety Assessments (A State-of-the-Art Report by a Group of Experts, 1980)	Les mouvements sismiques de référence du sol dans l'évaluation de la sûreté des installations nucléaires (Rapport sur l'état des connaissances établi par un Groupe d'experts, 1980)

£7.00 US$16.00 F64,00

Nuclear Safety Research in the OECD Area. The Response to the Three Mile Island Accident (1980)	Les recherches en matière de sûreté nucléaire dans les pays de l'OCDE. L'adaptation des programmes à la suite de l'accident de Three Mile Island (1980)

£3.20 US$8.00 F32,00

Safety Aspects of Fuel Behaviour in Off-Normal and Accident Conditions (Proceedings of the Specialist Meeting, Espoo, Finland, 1980) (in preparation)	Considérations de sûreté relatives au comportement du combustible dans des conditions anormales et accidentelles (Compte rendu de la réunion de spécialistes, Espoo, Finlande, 1980) (en préparation)

£ US$ F

Safety of the Nuclear Fuel Cycle (A State-of-the-Art Report by a Group of Experts, 1981)	Sûreté du Cycle du Combustible Nucléaire (Rapport sur l'état des connaissances établi par un Groupe d'Experts, 1981)

£ 6.60 US$ 16.50 F 66,00

• • •

SCIENTIFIC INFORMATION

Neutron Physics and Nuclear Data for Reactors and other Applied Purposes
(Proceedings of the Harwell International Conference, 1978)

INFORMATION SCIENTIFIQUE

La physique neutronique et les données nucléaires pour les réacteurs et autres applications
(Compte rendu de la Conférence Internationale de Harwell, 1978)

£26.80 US$55.00 F220,00

Calculation of 3-Dimensional Rating Distributions in Operating Reactors
(Proceedings of the Paris Specialists' Meeting, 1979)

Calcul des distributions tridimensionnelles de densité de puissance dans les réacteurs en cours d'exploitation (Compte rendu de la Réunion de spécialistes de Paris, 1979)

£9.60 US$21.50 F86.00

Nuclear Data and Benchmarks for Reactor Shielding
(Proceedings of a Specialists' Meeting, Paris, 1980)

Données nucléaires et expériences repères en matière de protection des réacteurs
(Compte rendu d'une réunion de spécialistes, Paris, 1980)

£9.60 US$24.00 F96,00

LEGAL PUBLICATIONS

PUBLICATIONS JURIDIQUES

Convention on Third Party Liability in the Field of Nuclear Energy — incorporating the provisions of Additional Protocol of January 1964

Convention sur la responsabilité civile dans le domaine de l'énergie nucléaire — Texte incluant les dispositions du Protocole additionnel de janvier 1964

Free on request — Gratuit sur demande

Nuclear Legislation, Analytical Study: "Nuclear Third Party Liability" (revised version, 1976)

Législations nucléaires, étude analytique: "Responsabilité civile nucléaire" (version révisée, 1976)

£6.00 US$12.50 F50,00

Nuclear Legislation, Analytical Study: "Regulations governing the Transport of Radioactive Materials" (1980)

Législations nucléaires, étude analytique: "Réglementation relative au transport des matières radioactives" (1980)

£8.40 US$21.00 F84,00

Nuclear Law Bulletin
(Annual Subscription — two issues and supplements)

Bulletin de Droit Nucléaire
(Abonnement annuel — deux numéros et suppléments)

£5.60 US$12.50 F50,00

Index of the first twenty five issues of the Nuclear Law Bulletin

Index des vingt-cinq premiers numéros du Bulletin de Droit Nucléaire

Description of Licensing Systems and Inspection of Nuclear Installation (1980)

Description du régime d'autorisation et d'inspection des installations nucléaires (1980)

£7.60 US$19.00 F76,00

NEA Statute

Statuts de l'AEN

Free on request — Gratuit sur demande

• • •

OECD SALES AGENTS
DÉPOSITAIRES DES PUBLICATIONS DE L'OCDE

ARGENTINA – ARGENTINE
Carlos Hirsch S.R.L., Florida 165, 4° Piso (Galería Guemes)
1333 BUENOS AIRES, Tel. 33.1787.2391 y 30.7122

AUSTRALIA – AUSTRALIE
Australia and New Zealand Book Company Pty, Ltd.,
10 Aquatic Drive, Frenchs Forest, N.S.W. 2086
P.O. Box 459, BROOKVALE, N.S.W. 2100

AUSTRIA – AUTRICHE
OECD Publications and Information Center
4 Simrockstrasse 5300 BONN. Tel. (0228) 21.60.45
Local Agent/Agent local :
Gerold and Co., Graben 31, WIEN 1. Tel. 52.22.35

BELGIUM – BELGIQUE
LCLS
35, avenue de Stalingrad, 1000 BRUXELLES. Tel. 02.512.89.74

BRAZIL – BRÉSIL
Mestre Jou S.A., Rua Guaipa 518,
Caixa Postal 24090, 05089 SAO PAULO 10. Tel. 261.1920
Rua Senador Dantas 19 s/205-6, RIO DE JANEIRO GB.
Tel. 232.07.32

CANADA
Renouf Publishing Company Limited,
2182 St. Catherine Street West,
MONTRÉAL, Quebec H3H 1M7. Tel. (514)937.3519
522 West Hasting,
VANCOUVER, B.C. V6B 1L6. Tel. (604) 687.3320

DENMARK – DANEMARK
Munksgaard Export and Subscription Service
35, Nørre Søgade
DK 1370 KØBENHAVN K. Tel. +45.1.12.85.70

FINLAND – FINLANDE
Akateeminen Kirjakauppa
Keskuskatu 1, 00100 HELSINKI 10. Tel. 65.11.22

FRANCE
Bureau des Publications de l'OCDE,
2 rue André-Pascal, 75775 PARIS CEDEX 16. Tel. (1) 524.81.67
Principal correspondant :
13602 AIX-EN-PROVENCE : Librairie de l'Université.
Tel. 26.18.08

GERMANY – ALLEMAGNE
OECD Publications and Information Center
4 Simrockstrasse 5300 BONN Tel. (0228) 21.60.45

GREECE – GRÈCE
Librairie Kauffmann, 28 rue du Stade,
ATHÈNES 132. Tel. 322.21.60

HONG-KONG
Government Information Services,
Sales and Publications Office, Baskerville House, 2nd floor,
13 Duddell Street, Central. Tel. 5.214375

ICELAND – ISLANDE
Snaebjörn Jönsson and Co., h.f.,
Hafnarstraeti 4 and 9, P.O.B. 1131, REYKJAVIK.
Tel. 13133/14281/11936

INDIA – INDE
Oxford Book and Stationery Co. :
NEW DELHI, Scindia House. Tel. 45896
CALCUTTA, 17 Park Street. Tel. 240832

INDONESIA – INDONÉSIE
PDIN-LIPI, P.O. Box 3065/JKT., JAKARTA, Tel. 583467

IRELAND – IRLANDE
TDC Publishers – Library Suppliers
12 North Frederick Street, DUBLIN 1 Tel. 744835-749677

ITALY – ITALIE
Libreria Commissionaria Sansoni :
Via Lamarmora 45, 50121 FIRENZE. Tel. 579751
Via Bartolini 29, 20155 MILANO. Tel. 365083
Sub-depositari :
Editrice e Libreria Herder,
Piazza Montecitorio 120, 00 186 ROMA. Tel. 6794628
Libreria Hoepli, Via Hoepli 5, 20121 MILANO. Tel. 865446
Libreria Lattes, Via Garibaldi 3, 10122 TORINO. Tel. 519274
La diffusione delle edizioni OCSE è inoltre assicurata dalle migliori librerie nelle città più importanti.

JAPAN – JAPON
OECD Publications and Information Center,
Landic Akasaka Bldg., 2-3-4 Akasaka,
Minato-ku, TOKYO 107 Tel. 586.2016

KOREA – CORÉE
Pan Korea Book Corporation,
P.O. Box n° 101 Kwangwhamun, SÉOUL. Tel. 72.7369

LEBANON – LIBAN
Documenta Scientifica/Redico,
Edison Building, Bliss Street, P.O. Box 5641, BEIRUT.
Tel. 354429 – 344425

MALAYSIA – MALAISIE
and/et **SINGAPORE - SINGAPOUR**
University of Malaysia Co-operative Bookshop Ltd.
P.O. Box 1127, Jalan Pantai Baru
KUALA LUMPUR. Tel. 51425, 54058, 54361

THE NETHERLANDS – PAYS-BAS
Staatsuitgeverij
Verzendboekhandel Chr. Plantijnnstraat
S-GRAVENAGE. Tel. nr. 070.789911
Voor bestellingen: Tel. 070.789208

NEW ZEALAND – NOUVELLE-ZÉLANDE
Publications Section,
Government Printing Office,
WELLINGTON: Walter Street. Tel. 847.679
Mulgrave Street, Private Bag. Tel. 737.320
World Trade Building, Cubacade, Cuba Street. Tel. 849.572
AUCKLAND: Hannaford Burton Building,
Rutland Street, Private Bag. Tel. 32.919
CHRISTCHURCH: 159 Hereford Street, Private Bag. Tel. 797.142
HAMILTON: Alexandra Street, P.O. Box 857. Tel. 80.103
DUNEDIN: T & G Building, Princes Street, P.O. Box 1104.
Tel. 778.294

NORWAY – NORVÈGE
J.G. TANUM A/S Karl Johansgate 43
P.O. Box 1177 Sentrum OSLO 1. Tel. (02) 80.12.60

PAKISTAN
Mirza Book Agency, 65 Shahrah Quaid-E-Azam, LAHORE 3.
Tel. 66839

PHILIPPINES
National Book Store, Inc.
Library Services Division, P.O. Box 1934, MANILA.
Tel. Nos. 49.43.06 to 09, 40.53.45, 49.45.12

PORTUGAL
Livraria Portugal, Rua do Carmo 70-74,
1117 LISBOA CODEX. Tel. 360582/3

SPAIN – ESPAGNE
Mundi-Prensa Libros, S.A.
Castello 37, Apartado 1223, MADRID-1. Tel. 275.46.55
Libreria Bastinos, Pelayo 52, BARCELONA 1. Tel. 222.06.00

SWEDEN – SUÈDE
AB CE Fritzes Kungl Hovbokhandel,
Box 16 356, S 103 27 STH, Regeringsgatan 12,
DS STOCKHOLM. Tel. 08/23.89.00

SWITZERLAND – SUISSE
OECD Publications and Information Center
4 Simrockstrasse 5300 BONN. Tel. (0228) 21.60.45
Local Agents/Agents locaux
Librairie Payot, 6 rue Grenus, 1211 GENÈVE 11. Tel. 022.31.89.50
Freihofer A.G., Weinbergstr. 109, CH-8006 ZÜRICH.
Tel. 01.3634282

TAIWAN – FORMOSE
National Book Company,
84-5 Sing Sung South Rd, Sec. 3, TAIPEI 107. Tel. 321.0698

THAILAND – THAILANDE
Suksit Siam Co., Ltd., 1715 Rama IV Rd,
Samyan, BANGKOK 5. Tel. 2511630

UNITED KINGDOM – ROYAUME-UNI
H.M. Stationery Office, P.O.B. 569,
LONDON SE1 9NH. Tel. 01.928.6977, Ext. 410 or
49 High Holborn, LONDON WC1V 6 HB (personal callers)
Branches at: EDINBURGH, BIRMINGHAM, BRISTOL,
MANCHESTER, CARDIFF, BELFAST.

UNITED STATES OF AMERICA – ÉTATS-UNIS
OECD Publications and Information Center, Suite 1207,
1750 Pennsylvania Ave., N.W. WASHINGTON D.C.20006.
Tel. (202) 724.1857

VENEZUELA
Libreria del Este, Avda. F. Miranda 52, Edificio Galipan,
CARACAS 106. Tel. 32.23.01/33.26.04/33.24.73

YUGOSLAVIA – YOUGOSLAVIE
Jugoslovenska Knjiga, Terazije 27, P.O.B. 36, BEOGRAD.
Tel. 621.992

Les commandes provenant de pays où l'OCDE n'a pas encore désigné de dépositaire peuvent être adressées à :
OCDE, Bureau des Publications, 2, rue André-Pascal, 75775 PARIS CEDEX 16.

Orders and inquiries from countries where sales agents have not yet been appointed may be sent to:
OECD, Publications Office, 2 rue André-Pascal, 75775 PARIS CEDEX 16.

PUBLICATIONS DE L'OCDE, 2, rue André-Pascal, 75775 PARIS CEDEX 16 - N° 41986 1981
IMPRIMÉ EN FRANCE
1500/SH (66 81 09 3) ISBN 92-64-02186-8